life on mars

life on mars

what to know before we go

david a. weintraub

princeton university press

princeton and oxford

Published by Princeton University Press,
41 William Street, Princeton, New Jersey 08540

In the United Kingdom: Princeton University Press,
6 Oxford Street, Woodstock, Oxfordshire OX20 1TR

press.princeton.edu

Jacket image: View toward "Vera Rubin Ridge" on
Mount Sharp, Mars / NASA/JPL-Caltech/MSSS

ISBN 978-0-691-18053-3

Library of Congress Control Number 2018931019

British Library Cataloging-in-Publication Data is available

This book has been composed in Odibee Sans & Sabon Next LT Pro

Printed on acid-free paper ∞

Printed in the United States of America

1 3 5 7 9 10 8 6 4 2

contents

For Caren Levy Burgess & Alan Burgess
& My West Coast parents, Dot and Gerry Levy

So much nonsense has been written about the planet . . . that it is easy to forget that Mars is still an object of serious scientific investigation.

—CANADIAN ASTRONOMER PETER M. MILLMAN, "IS THERE VEGETATION ON MARS," *THE SKY*, 3, 10–11 (1939)

life on mars

1

---

why mars matters

Are we alone in the universe? Earth might be an oasis of life, the only place in the universe where living beings of any kind exist. On the other hand, life might be as common across the universe as the hundreds of billions of stars and planets that populate it. If life is common, if the genesis of life is fairly easy given the right environment and the necessary elemental materials, some form of life might exist right next door, on Mars, and if life were discovered on Mars that is of an independent origin than life on Earth, we could safely predict that life is common throughout the universe. Such a discovery would be extraordinary. Mars Matters.

Mars has always attracted the attention of sky watchers on Earth, whether as the Greek (Ares) or Roman (Mars) or Babylonian (Nirgal) or Hindu (Mangala or Angaraka) God of War, or as the Chinese (Huo Hsing) or Japanese (Kasei) Fire star. The Incas named this planet Auqakuh; in ancient Sumer, it was called Simud; in ancient Hebrew, Ma'adim. Everywhere and for all of remembered history, Mars always had a name. We have been watching it for as long as we've been looking up into the heavens. As a planet (a wandering star, in the vernacular of the ancient Greeks), Mars stood out as a special object in the sky, comparable in brightness only to Venus, Jupiter, and Saturn, but even without a telescope, Mars is more colorful in the nighttime sky than the other planets, appearing red in color much of the time. Perhaps that is the lure of Mars. Perhaps the appeal of Mars in an ancient sky full of gods, in a celestial tapestry of myths, led us to imagine Mars as a place more special than all the other places we might visit in our imaginations.

We have been attracted to the idea of life on Mars for a very long time, by turns led and misled by our desires and imaginations. Thousands of years of human history, in which humans across all cultures invested the bright red planet with enormous mythological importance, combined with the medieval and Renaissance-era expectation that almost all other worlds should be inhabited, may have led astronomers to expect to find that Mars was Earthlike, and so they may have found what they wanted to find. The picture of Mars that emerged after the invention of the telescope revealed that the fourth planet from the Sun shares many life-critical similarities with Earth. With that realization, it was natural for astronomers to draw the conclusion that Mars must also be capable of hosting living things. As we approach a time when we might colonize Mars, we need to understand the burden of historical expectations regarding life on Mars that we all shoulder, because the history of discoveries about Mars made over the last four hundred years motivates today's scientists as they explore and study the red planet.

Does life exist on Mars? Maybe. Are the Martians Little Green Men? Not likely. Could primitive microorganisms survive on Mars, living in subsurface reservoirs of liquid water? Yes. Long ago, could spores have been transferred via a large impact event from Mars to Earth or from Earth to Mars? Very possibly.

We know that the six most important elemental building blocks of life—carbon, oxygen, nitrogen, hydrogen, phosphorous, and sulfur—exist virtually everywhere in the universe. The backbone of chemical life as we know it is carbon, which is abundant on Mars. Mars also has plenty of nitrogen and phosphorous, which are necessary components of both amino acids and DNA. We know Mars has water, which is made of hydrogen and oxygen, so those two elements, both together as water and separately for other chemical processes, are readily available. Sulfur, which occurs in all sugars, proteins, and nucleic acids, is also abundant on Mars. Chemically, at least, Mars has all the right stuff for the origin and survival of chemically based life. In addition, Mars has spent part of the last 4.5 billion years since the birth of the Sun in the Goldilocks zone of our solar system, where temperatures, pressures, and densities are just right to allow liquid water to exist on or just below the Martian surface for at least part of Mars's annual cycle of seasons.

In principle, Mars could therefore harbor life, whether as a birthplace for living things or as a nurturing environment for life-forms that might have been deposited there. Furthermore, Mars may be a model for helping us understand the likelihood that life could exist on any of the many exoplanets recently discovered by astronomers that lie in the Goldilocks zones around their host stars.

Some scientists have speculated that life in our own solar system may have started on Mars and later been accidentally exported to Earth when a large asteroid collided with Mars and splashed Martian rocks into space. Alternatively, though dynamically more difficult, life may have started on Earth and, as the result of a large terrestrial impact event, been transported to Mars, in which case Mars could function as a nurturing environment for life-forms that might have been deposited there after forming elsewhere.

Earth and Mars formed at about the same time, almost 4.5 billion years ago, in a swirling disk of gas and dust that orbited the newborn Sun. After suffering through the traumatic early stages of formation, in which large asteroids and myriad comets likely continually crashed into their surfaces for several hundred million years, the solar system stabilized. As soon as the planets cooled enough for solid land to begin to form, both Earth and Mars almost certainly also had liquid water pooling on their surfaces.

On at least one of these two planets, life made an appearance fairly soon after these primordial formation events. Australian geologist Allen Nutman and his scientific team recently pushed the time for the oldest known living things on Earth all the way back to 3.7 billion years ago. In a rock formation in Isua, Greenland, they found layered structures that prove that the rock itself is a stromatolite.[1] Such rocks form as colonies of microorganisms deposit layers of minerals as they grow. From Nutman's work, we therefore know that stromatolites thrived in shallow seas on Earth when our planet was only 800 million years old. Incredibly, life must have taken root on Earth very quickly after it formed. Equally certainly, Mars was also

Figure 1.1. Living marine stromatolites in Hamlin Pool, Australia. See Plate 1. Image courtesy of Kristina D. C. Hoepper/Creative Commons at https://www.flickr.com/photos/4nitsirk/11902636365

warm and wet when it too was only 800 million years old. If so, then life could have formed or grabbed hold on Mars at about the same time. Thus, the existence of stromatolites on a young Earth strongly suggests that a young Mars, too, could have had colonies of primitive living things in its shallow lakes.

Without question, Mars is the closest place in the universe where we plausibly might find extraterrestrial life. For centuries, astronomers repeatedly have claimed to have discovered evidence suggesting Mars harbors life; however, to date all of these findings have either been disproven or become highly disputed. Where does this leave us? Today, we lack anything resembling a scientific consensus regarding how to answer the question, "Does or did life exist on Mars?" The existence of so many varied and controversial claims for life on Mars suggests the tantalizing possibility that life once thrived there or even exists on Mars today; yet, we cannot point to any one piece of unassailable evidence for once or current life on Mars. The jury is still out; it needs more evidence.

The discovery of extraterrestrial life on Mars would rank among the most profound and important discoveries ever made in the history of science. Such a discovery would also raise enormous ethical and moral concerns. If scientists conclude that life exists on Mars, then the debate as to whether we should colonize Mars, knowing that it is already inhabited, could become one of the most important questions facing us in the mid-twenty-first century. Does humanity have an inalienable right to potentially disrupt life on another world simply because we have the technological ability to transport members of our species across interplanetary space? Some ethicists would argue that if Mars is home to nothing more biologically advanced than a few colonies of microbes, we should feel free to colonize the red planet, while if we found multi-celled creatures, we should leave them alone.

How soon might humans set foot on Mars? As instructed by the NASA Authorization Act of 2010 and the U.S. National Space Policy, also issued in 2010, NASA is developing the capabilities for sending humans to Mars and returning them safely to Earth by the 2030s. The time frame for reaching Mars in these plans is likely overly optimistic, and NASA is gradually scaling back expectations; nevertheless, we are planning to send astronauts to Mars within the lifetimes of many of us. Current plans include a first phase of exploration in the vicinity of the Moon, including building a spaceport in lunar orbit, which would be NASA's gateway to deep space—that being targets well beyond the Moon.

NASA's human missions to Mars (and the Moon) will be launched from the Kennedy Space Center in Florida, where an advanced tracking system designed to support the goal of sending humans beyond the Moon is already nearing completion. The Space Launch System (SLS), if completed, will be about 20 percent more powerful than the Saturn V rocket that supported the Apollo program for exploration of the Moon (the Saturn V could lift 135 tons into orbit) and will be built using the same, time-tested rocket technology developed for the Space Shuttle program.

The tremendous power of the SLS may eventually send astronauts to Mars in the Orion Multi-Purpose Crew Vehicle. The Orion would provide living space for the astronauts during the 16-month-long round-trip journey to Mars. The first SLS vehicle, known as Block 1, which has a completion goal of 2018, will have a lift capability of 77 tons. The plan for the first SLS mission is to launch a spacecraft to an orbit past the Moon and then return that vehicle to Earth. The next design phase of SLS, Block 1B, is intended to add a more powerful upper stage, giving SLS a planned lift capability of 115 tons. NASA intends to use this configuration to send astronauts well beyond the Moon, perhaps to the vicinity of a near-Earth asteroid. The third design phase of SLS, Block 2, includes plans to replace the five rocket boosters on Block 1 with solid or liquid propellant boosters that, in design, are intended to have a 143-ton lift capability. Current estimates of the planned final configuration of the SLS that will launch astronauts to Mars are that this vehicle will weigh 6.5 million pounds, comparable to 10 fully loaded 747 jets; provide 9.2 million pounds of thrust at liftoff, equivalent to more than 208,000 Corvette engines; and stand 365 feet high, taller than a thirty-story building.

The first, unmanned, two-orbit test flight of Orion was carried out in December 2014. The first integrated launch and flight of the SLS rocket and the Orion spacecraft to a point beyond the Moon, known as Exploration Mission–1, is now scheduled, without an astronaut crew, for 2019.[2] The first crewed flight of astronauts on Orion, Exploration Mission–2, was scheduled for 2021, but likely will be delayed. The second phase of exploration, which includes plans for an eventual trip to Mars, will begin in the late 2020s with a planned one-year crewed mission to the lunar spaceport. By the 2030s, NASA intends to have tested all the systems and capabilities of Orion necessary to carry astronauts and life-critical cargo to Mars's orbit and then return them to Earth.

Sending astronauts to Mars, landing them on the surface, keeping them alive, and then lifting them back off the Martian surface and returning them safely home remain well beyond NASA's current capabilities. The downward pull of gravity at the surface of Mars is almost 2.5 times greater than the pull of gravity at the surface of the Moon. As a result, safely landing astronauts on Mars requires retrorockets or using some other lander design that will slow the downward acceleration of astronauts toward the surface of Mars. For the same reason, blasting back off of Mars will be a much greater technological challenge than was returning astronauts from the surface of the Moon. Getting to and from Mars, of course, is only part of the problem of living on Mars, and NASA is already working on imaginative plans for building a Mars colony.

NASA, however, is no longer the only player in the exploration of space and the race to Mars. PayPal founder and entrepreneur Elon Musk made clear, when he founded his SpaceX corporation in 2002, that his goal was to establish a human colony on Mars. Already, SpaceX has successfully delivered cargo to the Space Station in its Dragon spacecraft, which one day is intended to ferry astronauts to the Space Station and then carry them farther into space. SpaceX's current rocket, the two-stage Falcon 9, has the thrust of five 747s at full power and can lift 28 tons into orbit. In December 2015, SpaceX successfully demonstrated that the first stage can be safely landed back on Earth for reuse, and in May 2017, reused a first stage in a second rocket launch for the first time. SpaceX is now working toward the launch of the much more powerful Falcon Heavy, which is supposed to be capable of lifting 55 tons into orbit.

In June 2016, in an interview with the *Washington Post*, Musk first offered hints about his audacious plans for sending his first unmanned flight to Mars in 2018. Then in September 2016, he explained those plans in more detail at the International Astronautical Congress (IAC) in Guadalajara, Mexico. Two years later, he updated those plans in a presentation to the IAC in Adelaide, Australia, where he discussed making the human race "a multiplanet species." Musk's planned launch vehicle, which has had several names, including the Interplanetary Transport System (ITS) and the BFR ("B" for "big" and "R" for "rocket"), will be powered by thirty-one Raptor rocket engines, with a liftoff thrust of 5,400 tons (almost 11 million pounds), capable of lifting 150 tons into orbit. The BFR will replace all the previous SpaceX rockets and spacecraft (Falcon 9, Falcon Heavy, Dragon). According to current plans, the Raptors, which are still in the design phase by SpaceX, will use carbon-fiber tanks holding (separately) liquid methane and liquid oxygen as fuel and will be able to refuel in space, which would enable the BFR to then take all 150 lifted tons all the way to Mars. In 2022, Musk intends to use two ITS launches to send to and land two cargo payloads on Mars in Dragon spacecraft. If successive launches follow the company's extremely aggressive (many would say unrealistic) time line, SpaceX intends to launch four rockets to Mars in 2024; two would ferry additional cargo and two would carry human crews with up to one hundred adventurers in each, who would arrive at and land on Mars, establish a colony, construct a propellant production depot, and find a supply of water. Sci-fi fans might note the similarity of Musk's plans to those described in Kim Stanley Robinson's 1990s award-winning trilogy *Red Mars*, *Green Mars*, and *Blue Mars*, in which the First Hundred colonists were launched to the red planet in 2026.

Over the next forty years, Musk wants to shuttle as many as a million colonists to Mars and to begin manipulating the Martian climate to make it more like that of Earth, a concept referred to as terraforming. He claims that his Martians will be able to sustain themselves on Mars and also come home again, as his rockets will make regular roundtrips from Earth to Mars and back again. The assumption that colonists will ever get back to Earth, however, depends first on their ability to survive the harsh radiation environment of space and on the surface of Mars, and then on the ability of SpaceX to use solar power to manufacture methane and oxygen fuels on Mars (from subsurface and atmospheric reservoirs of water and carbon dioxide) for return trips. Musk's audacious plans also may require an infusion of tens to hundreds of billions of dollars for development, which is beyond his personal ability to fund.

Another tech billionaire, Amazon founder Jeffrey Bezos, is also building rockets through his venture Blue Origin and has his own plans for sending colonists to Mars. In 2016, Blue Origin successfully launched and landed its first rocket, the New Shepard, named after the United States' first astronaut Alan Shepard, on a suborbital flight. Blue Origin also is developing a more powerful rocket called New Glenn, named for the United States' first astronaut to orbit Earth, John Glenn, which it intends to launch from a massive facility it has under construction at the Kennedy Space Center's Exploration Park at Cape Canaveral, Florida. Plans for New Glenn, which may debut in 2020, include a reusable first-stage rocket as part of a three-stage, 350-foot-tall launch vehicle that will burn liquid hydrogen and liquid oxygen. Bezos expects his project to take many decades, rather than a single decade. First, Bezos expects his Blue Origin rockets to launch satellites and cargo, with a goal of delivering the equipment to the Moon that would be necessary to support a human colony there. Then he intends to place millions of humans into space, where they will work in near-Earth orbit. Only then will he set his sights on placing colonists on both the Moon and Mars. According to Bezos, "I think that if you go to the Moon first, and make the Moon your home, then you can get to Mars more easily."[3]

NASA may also have additional competition from a private Dutch group, Mars One.[4] Founded in 2011 by Bas Lansdorp and Arno Wielders, Mars One intends to launch an unmanned mission to Mars in 2020, a first crew to Mars in 2031, and send a second crew in 2033. Mars One began astronaut selections in 2013 and intends to select their first crews in 2017, at which time they will begin training for their one-way trip to Mars. In contrast to NASA, SpaceX, and Blue Origin, Mars One is not designing or manufacturing rockets, launch systems, landing modules, life support units, or rovers. Instead, they intend to purchase everything they need from established aerospace companies. Whether Mars One can actually buy the mission hardware they need remains to be seen.

One other player has recently announced plans for colonizing Mars. In February 2017, at the World Government Summit in Dubai, Sheikh Mohammed bin Rashid Al Maktoum, ruler of Dubai and vice president of the United Arab Emirates, announced that the UAE planned to build a city on Mars within one hundred years.[5] The UAE's Mars 2117 Project is only a concept for now; however, the UAE created its own space agency in 2014 and has plans to send an unmanned probe to Mars by 2021, coinciding with the fiftieth anniversary of the UAE gaining political independence from Great Britain in 1971.[6]

The reason so many of us are so interested in Mars is that life on Mars is possible, whether for us in the future or for native Martians. What if Mars harbors life today? What if astronauts establish a human colony on Mars in the twenty-first century? Will we bring death and destruction to Mars, as the first European colonists did to the New World, when they brought smallpox, measles, whooping cough, bubonic plague, and dysentery to a world that lacked the ability to fend off those attackers? They also brought horses and pigs that often outcompeted indigenous wildlife species for survival. Together, the Old World diseases and animals wreaked havoc on the biota of the New World. Humanity also does not have a good track record in taking care of remote wilderness areas. The ecosystems of the Arctic, Antarctic, and Amazon are all threatened by the encroachment of human civilization, by hunting and by global warming. If we cannot find the collective will to help the polar bears, penguins, and giant otters survive on our own planet, will we do anything to help ensure the survival of microscopic Martians?

Do microscopic Martians even matter? Yes. A second genesis, life that began completely independently of terrestrial origins, might have occurred there. Even if life on Mars is limited to bacterial-sized beings, buried underground or hiding deep in a crevice where they are protected from dangerous ultraviolet radiation and cosmic rays and where they can find water, those microscopic Martians would be astoundingly important to our understanding of life in the universe. Life on Mars that is independent of life on Earth would send us a clear message about exobiology: life happens anywhere and everywhere that conditions allow. Alternatively, if we find microscopic life that is DNA-based, we also receive an enormously important message about exobiology and clues about our distant, evolutionary past: life is easily transported across interplanetary space. Once life gets started, it spreads, and thus, whether we are Martians or the Martians are us, we're all related. Finally, if we find that Mars is barren and sterile, without even microscopic Martians, we will know that we are more alone in the solar system and perhaps in the galaxy and universe than many of us currently assume. Whatever the answer, the answer matters. Mars matters.

As the possibility of travel to Mars draws closer, we have an urgent scientific imperative to determine whether life exists on our planetary neighbor. Putting astronauts into Mars's orbit creates very little risk for contaminating Mars. Landing habitat modules and astronauts on Mars and attempting to build a colony on Mars, however, could inadvertently destroy any life that might exist on Mars before we have a chance to fully explore the red planet and discover whether life exists there.

The scientific detective story that follows traces the many attempts to identify life on Mars from the seventeenth century until now. As we consider these claims and discoveries, we might think about whether some caution is in order before we begin to colonize Mars. Perhaps our decision as to whether to colonize Mars should not be left strictly to politicians, professional astronauts and astronomers, space enthusiasts, and deep-pocketed venture capitalists. We all should better understand Mars, and we all should participate in this public debate.

2

―――

martians?

The contours of our collective thinking about Mars have been shaped, ever so slowly, by two centuries of gradually improving astronomical observations of the red planet. From the late seventeenth century into the mid-nineteenth century, several generations of astronomers used their increasingly powerful telescopes to map the surface of Mars with ever-improving clarity. Without a doubt, astronomers of recent centuries believed they were discovering clues at the far ends of their telescopes that revealed Mars was likely the home of Martians.

Once astronomers had determined that Mars shared intriguing similarities with Earth, the international community of astronomers undertook a decades-long effort to map the Martian globe. In the nineteenth century, they used then-new measuring tools to discover water in the Martian atmosphere. Or at least they thought they had done so. With maps of Mars that included surface features astronomers identified as Earthlike, including continents and oceans, and armed with the belief that they had identified evidence of water that cycled between the atmosphere, oceans, and polar ice caps, astronomers convinced themselves and the public that Mars was, in all ways, just like Earth.

In the late nineteenth century, two astronomers, first the Italian Giovanni Schiaparelli and then the American Percival Lowell, constructed a new vision of Mars that reshaped our relationship with our planetary neighbor. In 1878, Schiaparelli first identified *canali* on Mars. In Italian, *canale* is the word for a channel, and can be used to refer to a human-made canal in Venice, the English Channel, or a mountain gully. In the 1890s, Lowell took Schiaparelli's canali one step further when he announced that the canali were artificially constructed canals engineered by supremely intelligent Martian engineers. Our continuing twentieth- and twenty-first-century searches for evidence of life on Mars are the indirect legacy of Schiaparelli's discoveries and of Lowell's unfettered imagination and indefatigable efforts to promote his ideas about Mars to the public.

martian invaders

Not surprisingly, in the same decade of the 1890s when Lowell was barnstorming the country and drawing the public toward his ideas about an ancient, intelligent civilization on Mars, H. G. Wells published his *War of the Worlds*, in which a vanguard of Martian invaders attack Earth.

"We know now that in the early years of the twentieth century this world was being watched closely by intelligences greater than man's and yet as mortal as his own."[1] These were the first words spoken by Orson Welles, as the Columbia Broadcasting System's Sunday evening *Mercury Theater on the Air* broadcast began on October 30, 1938. Orson Welles was putting on the air a modernized adaptation of H. G. Wells's science fiction novel, originally penned in 1898. "We know now," he continued, "that as human beings busied themselves about their various concerns they were scrutinized and studied, perhaps almost as narrowly as a man with a microscope might scrutinize the transient creatures that swarm and multiply in a drop of water." The dramatization segues to an announcer offering a weather forecast for the northeastern United States, after which listeners are entertained by the music of a dance band playing live, they are told, in the Meridian Room of the Park Plaza Hotel in New York City. Suddenly, a second announcer interrupts the music to bring the audience a special news bulletin from Intercontinental Radio News. The bulletin advises listeners that Professor Farrell, of the Mount Jennings Observatory in Illinois, had observed "several explosions of incandescent gas, occurring at regular intervals on the planet Mars."

We learn from Princeton professor of astronomy Richard Pierson, who confirmed the astounding observations of his colleague, that astronomers had been observing Mars but were unable to explain these eruptions. Soon thereafter in *Mercury Theater*'s broadcast, an enormous, flaming meteorite fell through Earth's atmosphere and impacted on the grounds of the Wilmuth family farm in Grovers Mill, New Jersey. The object that struck Earth on the Wilmuths' farm, however, was not a meteorite. It was a huge, smooth, metallic cylinder, from which, soon after impact, witnesses heard clanking sounds. Aghast listeners learned from radio commentator Carl Phillips, who, in the magically compressed time of the *Mercury Theater* broadcast, had been dispatched to the impact site, that "something's wriggling out of the shadow like a gray snake. Now it's another one, and another. They look like tentacles to me. There, I can see the thing's body. It's large, large as a bear and it glistens like wet leather. But that face, it . . . Ladies and gentlemen, it's indescribable. I can hardly force myself to keep looking at it. The eyes are black and gleam like a serpent. The mouth is V-shaped with saliva dripping from its rimless lips that seem to quiver and pulsate."

The Martian invaders began throwing flames at the Earthlings who had gathered on Wilmuths' farm, burning everything within range of their heat rays. After Phillips's broadcast was cut short by the sound of explosions, a radio announcer continued to broadcast to frightened listeners. "Ladies and gentlemen," he intoned, "I have a grave announcement to make. Incredible as it may seem, both the observations of science and the evidence of our eyes lead to the inescapable assumption that those strange beings who landed in the Jersey farmlands tonight are the vanguard of an invading army from the planet Mars."

According to some media reports, in the immediate aftermath of the *War of the Worlds* radio broadcast, much of the citizenry of the United States began to panic over an imagined invasion of Earth by an army of Martians. Historian Richard Ketchum reports, in his book *The Borrowed Years*, "Terrified New Yorkers began leaving their apartments, some heading for city parks, others in automobiles jamming Riverside Drive in an effort to get out of town, still others crowding rail and bus terminals. San Franciscans got the impression that New York City was being destroyed. . . . in Indianapolis, [where] a woman rushed into a church, screaming, 'New York destroyed . . . It's the end of the world! You might as well go home to die.'"[2] According to one report, "Panic broke out across the country. In New Jersey, terrified civilians jammed highways seeking to escape the alien marauders. People begged police for gas masks to save them from the toxic gas and asked electric companies to turn off the power so that the Martians wouldn't see their lights."[3]

This oft-told version of the history of the *War of the Worlds* event emerged from a research investigation by Princeton University professor of psychology Hadley Cantril, published in 1940 as *The Invasion from Mars: A Study in the Psychology of Panic*. Disappointingly perhaps, for those who think our grandparents' generation was full of uneducated and gullible fools, Cantril's study was seriously flawed. Panic did not break out across the country since most people were not listening to their radios, and most of those listening were not tuned in to the *Mercury Theater* broadcast. What little hysteria did occur was mostly limited to New Jersey and the vicinity of New York City. A. Brad Schwartz, in his recent, more complete study of the *War of the Worlds* phenomenon, published in 2015 as *Broadcast Hysteria: Orson Welles's 'War of the Worlds' and the Art of Fake News*, corrects the earlier understanding of what happened during and immediately after listeners heard about the Martian invasion on their radios. According to Schwartz, Cantril drew incorrect conclusions about the impact of the *War of the Worlds* broadcast because his research team "deliberately oversampled people frightened by the broadcast, ignored survey data from listeners who knew it was fiction, and only interviewed listeners in New Jersey—where all accounts agree that the panic was most intense."[4] "The vast majority of listeners," Schwartz writes, "understood the broadcast correctly, and those few who were frightened did not passively accept what came to them over the airwaves. They often tried to verify the information in any way they could." Whatever little panic occurred "only began when some listeners passed the fake news on to unsuspecting others, spreading their fear and confusion. This behavior was not the mass headlong flight from reality that the word 'panic' implies."[5]

Nevertheless, the fact that any listeners were ready to believe this radio dramatization of H. G. Wells's science fiction story about Mars was true tells us almost everything we need to know about the relationship of Mars to human thought and ideas in the first third of the twentieth century. By the 1930s, when Orson Welles broadcast *War of the Worlds* on *Mercury Theater on the Air*, many of his listeners had been reading about Martians for decades. George du Maurier penned his *The Martian* (1897) contemporaneously with H. G. Wells's *War of the Worlds*. In du Maurier's gothic science fiction story, a Martian named Martia inhabits the body of a human. Martia came to Earth from Mars as a shooting star and spent most of a century inhabiting various living things before eventually choosing to incarnate as Barty Josselin, an English literary genius. Many other novels about Martians appeared, including *A Prophetic Romance: Mars to Earth* (John McCoy, 1896), *Edison's Conquest of Mars* (Garret P. Serviss, 1898), *Lieut. Gullivar Jones: His Vacation* (Edwin Lester Arnold, 1905), and *Doctor Omega* (Arnould Galopin, 1906). Even C. S. Lewis wrote about Martians, in his *Out of the Silent Planet* (1938).

Less than two decades after *The Martian* appeared, Edgar Rice Burroughs serialized his first stories about John Carter on Mars as *Under the Moons of Mars* (1912). Burroughs's enormously popular hero Carter was a captain in the Confederate Army who finds himself mysteriously transported from Earth to Barsoom, the fictional name Burroughs used for Mars. Burroughs published a dozen more of these novels over thirty years, beginning with *A Princess of Mars* (1917), with his Martian civilization closely resembling Lowell's ideas for Mars, in the sense that Burroughs's Mars is old and dry and the civilization he depicts on Mars, which includes mad scientists, green and red Martian warriors, and the chaste princess with whom John Carter falls in love, is dying.

At about the same time that John Carter was falling in love with a Martian princess, *A Trip to Mars* appeared in movie theaters—twice. The Edison Production Company released the first *A Trip to Mars* in

1910. This five-minute movie tells the story of a scientist who discovers a powder that reverses gravity. As the result of the chemical spill of the anti-gravity powder, the scientist is transported to Mars, where he has to escape from a dense forest of enormous trees, each of which is a monster with arms that attempts to capture our hero. The second *A Trip to Mars*, originally a Danish film titled *Himmelskibet* (*Heaven's Ship*), was released in 1918 as one of the first full-length (80-minute) silent motion pictures. This version of *A Trip to Mars* depicts a team of scientists who fly to Mars and encounter a peace-loving civilization of Martians. In 1913, in between the release of the two *A Trip to Mars* movies, United Kingdom Photoplays released the one-hour film *A Message from Mars*. This movie was based on a play by Richard Ganthoney, first produced in 1899, in which a Martian is sent to Earth to redeem himself by making an Earthling perform a good deed. In 1934, Buck Rogers battles to protect Earth in *An Interplanetary Battle with the Tiger Men of Mars*, and in the fifteen-episode 1938 serial, *Flash Gordon's Trip to Mars*, Flash Gordon struggles week after week against a deadly ray gun, placed on Mars by Ming the Merciless of Planet Mongo.

By the 1930s, many educated non-scientists who read popular magazines like *Scientific American*, *Popular Astronomy*, and *Science News*, believed in Martians. As for the non-readers, in addition to Buck Rogers and Flash Gordon, Hollywood offered cartoons with Martians (*Felix the Cat Flirts with Fate* [1926]; *Up to Mars* [1930]; *Oswald the Lucky Rabbit: Mars* [1930]; *Scrappy's Trip to Mars* [1937]; *Believe It or Else* [1939]; *Mars: A Fantasy Travelogue* [1943]; Popeye's *Rocket to Mars* [1946]). For those who thought much about Mars, life on Mars was not a matter of "if." The public debate about life on Mars was about the question "What is Martian life like?" rather than the more fundamental question "Do Martians exist?" The idea that Martians could be so advanced that they could launch fleets of spaceships with armed invaders across the tens of millions of miles of space that separate our two planets did not surprise or shock many of Orson Welles's listeners in 1938. Incredibly, less than a century later, we are the ones capable of launching a fleet of spaceships across the solar system and sending terrestrial invaders to Mars.

no lunatics or solarians

To some great thinkers of the ancient and medieval worlds, worlds other than Earth were full of living creatures, though they based their claims on little more than their philosophical preferences for not being alone.[6] The Greek philosopher Epicurus (341–270 BCE) argued that other worlds, full of living creatures, must exist, writing, "there is an infinite number of worlds, some like this world, others unlike it. . . . nobody can prove that in one sort of world there might not be contained . . . the seeds out of which animals and plants arise."[7] Six centuries later, Pseudo-Plutarch asserted, "the moon is terraneous, is inhabited as our earth is, and contains animals of a larger size and plants of a rarer beauty than our globe affords."[8]

Later, religious scholars would draw the same conclusion. The revered medieval Jewish authority Moses Maimonides (1135–1204 CE) advised others to "Consider how vast are the dimensions and how great the number of corporeal beings . . . The species of man is the least in comparison to the superior existents—I refer to the spheres and the stars."[9] Two hundred years later, Roman Catholic cardinal Nicolaus Cusanus (1401–1464 CE) wrote, "Life, as it exists on earth in the form of men, animals and plants, is to be found, let us suppose, in a higher form in the solar and stellar regions. . . . we will suppose that in every region there are inhabitants . . . It may be conjectured that in the area of the sun there exist solar beings, bright and enlightened denizens, and by nature more spiritual than such as may inhabit the moon."[10] This way of thinking peaked with the speculations of the Italian Dominican monk Giordano Bruno (1548–1600 CE), who in 1584 asserted that "There are innumerable suns, an infinite number of earths revolve around those suns," that "the fiery worlds are inhabited," and that "solar creatures exist." Bruno then has one of his characters ask, "Then the other worlds are inhabited like our own?" Another Bruno character provides the affirmative answer, "If not exactly as our own, and if not more nobly, at least no less inhabited and no less nobly."[11] On February 17, 1600, the Roman Inquisition burned Bruno at the stake in the Roman square known as Campo de' Fiore. While we do not know the exact reasons Bruno was condemned, and while his controversial views on extraterrestrial life were and are well known and were certainly among his troubles, he gave his inquisitors several other reasons for branding him a heretic: Bruno was accused of asserting that Christ was an unusually skilled magician who performed illusory miracles; of reading forbidden books; and of having unorthodox views on the Trinity, reincarnation, the (non)existence of Hell, and transubstantiation.[12]

By the time Galileo turned us into Peeping Toms, other than knowing that five other planets and the Sun and the Moon existed, we didn't know anything about what these distant celestial bodies were like. Every idea anyone had about other worlds, including life on the Sun, the Moon, or any of the planets, was pure speculation. Then, in 1610, using his improved version of Dutch spectacle maker Hans Lipperhey's new invention, the spyglass, Galileo showed astronomers how to part the curtains of the heavens. In the decades and centuries that followed, metaphysics—that is, physics explained on the basis of philosophical principles rather than on the basis of hard data—would slowly give way to knowledge about our planetary companions in orbit around the Sun, but our knowledge would also continue to be deeply colored by our expectations and philosophical biases.

When the telescope arrived on the scene, the principle of plenitude, an invention of the human intellect already more than a thousand years old that asserts that every possible form of existence occurs in the universe, offered a clear answer to Renaissance-era scholars who were asking the seemingly unanswerable question, Does life exist on other planets? Yes, insists the principle of plenitude! Extraterrestrial beings must exist because God's goodness demands that all worlds, including Venus and Mars, Jupiter and Saturn, the Moon and even the Sun, should be populated with intelligent, God-worshipping denizens. After all, according to the principle of plenitude, the worship of God by those beings is the very reason why God created those worlds. Armed with both the knowledge that extraterrestrial life must exist and Galileo's technological innovation, extraterrestrials could no longer use the great depths of space that separated their planets from Earth as a shield, behind which they could hide from astronomers. As a result, before human eyes peered at Mars through a telescope for the first time, we were predisposed to believe that life existed on our closest planetary neighbor.

The telescope, however, and the scholars born into a "modern" world that included the telescope, changed the rules. Natural philosophy and metaphysics gave way to data-driven science and experimental physics. In the seventeenth century, the measurements made by brilliant observers like Galileo Galilei and Robert Boyle and the calculations of mathematical geniuses like Johannes Kepler and Isaac Newton created new knowledge that pushed the old, stale ideas of metaphysics, ideas like Aristotle's geocentric universe and Ptolemy's epicycles, into the dustbin of history. Slowly but surely, serious astronomical observations, made with ever-improving telescopes by the great astronomers of the seventeenth century, like the Italian Giovanni Domenico Cassini and the Dutchman Christiaan Huygens, began to chase away our imaginary neighbors on the Moon, the Sun, and most of the other five planets. Astronomical knowledge began to displace the guesswork of earlier ages and the solar system gradually became a lonelier place for us humans. Eventually, Mars would be left as the only plausible locale in the known solar system where extraterrestrial life might exist. At that point, all the intellectual and emotional energy invested by astronomers in considering the possibility of life beyond Earth became focused on a single planet.

Life on the Moon? Even to ancients viewing the Moon with nothing more than their unaided eyes, the Moon has obvious dark patches surrounded by broad, lighter colored expanses. Beginning in ancient times, astronomers called the dark patches on the Moon seas, or maria, because they thought the seas and oceans on Earth, if viewed from space, would look similarly dark in comparison to their surrounding, sandy-colored continents. By virtue of their supposed similarity in color and shape to Earth's large basins of water, the lunar maria were also assumed to be filled with water. The lighter-colored expanses were presumably vast plains situated somewhat above lunar sea level. In addition to the seas and plains, Galileo discovered in some of his earliest telescopic observations that the Moon has mountains, the lunar highlands elevated thousands of feet above the rest of the surface, that cast long shadows across the lunar plains.

Together, the seas, mountains, and plains made the Moon appear Earthlike, and surely an Earthlike world would be inhabited. In the late eighteenth century, the great French biologist Georges Louis Leclerc, Comte de Buffon, offered a slew of arguments in favor of the existence of lunar inhabitants. At about the same time, the astronomer William Herschel, the eminent German-Englishman who discovered the planet Uranus, proved that some stars orbit other stars in what he called "binary star" systems and even mapped the shape of the Milky Way galaxy, wrote favorably about the likelihood that "Lunarians" exist.[13]

While the seas and mountains made the Moon appear Earthlike, the sharp boundary between the visual edge of the Moon, the lunar limb, and the darkness of space was unlike that of Earth. If the Moon, like Earth, had an atmosphere, then light from distant stars should twinkle when they were seen extremely close to the edge of the Moon, such that the starlight would pass through the Moon's atmosphere. On the other hand, if the Moon did not have an atmosphere, then starlight would be completely unaffected, no matter how close the star was to the Moon. In the early nineteenth century, the great German astronomer Friedrich Wilhelm Bessell made telescopic observations that demonstrated the sharpness of the transition from the lunar limb to the crisp darkness of space. His measurements of the absence of a fuzzy edge to the Moon convinced most astronomers that the Moon lacked an atmosphere. In a pamphlet published in 1874, Harvard professor of paleontology Louis Agassiz, one of the great scientists of his day and one of the founding members of the National Academy of Sciences,* compiled a lengthy list of reasons why the Moon had no atmosphere. If the Moon had an atmosphere, he reasoned, the atmosphere would scatter sunlight into the shadowed regions, making the shadows less dark. Furthermore, he argued, the absolute blackness of the shadows of the lunar mountains offers proof of the absence of a lunar atmosphere. This, he wrote, is "the first proof that the Moon has no appreciable atmosphere." He offered the crisp transition from Moon to background stars, observed by Bessell, as an additional reason that proved the Moon had no atmosphere.[14]

---

*Agassiz is also remembered, correctly, as an anti-Darwinist and as a racist who believed that his views of absolute white superiority had a scientific basis.

With no atmosphere, the life-on-the-Moon hypothesis lost credibility. Astronomers understood that the lunar "seas" were almost certainly not actually filled with water. Instead, they were just dark-colored landforms on the lunar surface. Furthermore, without an atmosphere the Moon would have no air for animals and plants to breathe. The conclusion was obvious: no atmosphere means no water or air. No water or air means no life. Life could not exist on the surface of the Moon. Thus, by the middle of the nineteenth century, astronomers recognized that, despite simple and deceptive appearances, our closest neighbor is decidedly un-Earthlike. Maybe Lunarians could survive underground, but with only a few exceptions from astronomers on the fringe, support for life on the Moon quickly faded after astronomers learned how to study the Moon carefully through their telescopes. With observations made with modern telescopes, from our explosive growth in understanding the structure and history of the Moon as a result of the 1960s-era Apollo missions that placed astronauts on the lunar surface, and more recently from twenty-first-century orbiter missions to the Moon, we know with as much certainty as can be had in science that the Moon is a biologically sterile world.

Life on the Sun? A few highly respected intellectuals speculated about the existence of Solarians, most famously the eighteenth-century German astronomer Johann Elert Bode, director of Berlin Observatory and founder of the astronomical almanac* *Berliner Astronomisches Jahrbuch*, which was published annually from 1776 until 1960. Most astronomers understood that the Sun was the source of light and heat for Earth. They speculated, further, that it might also do the same for the rest of the universe. As the giver of all that heat, they concluded correctly that the Sun must, itself, be scorchingly hot. While Bode thought his Solarians could endure the blinding brightness and intense heat of the Sun, few others took seriously the idea that a burning hot Sun might harbor life. Thus, in the increasingly scientific culture of astronomy of the nineteenth century, serious astronomers eliminated the Sun as a viable location for any form of extraterrestrial life.

---

*An astronomical almanac contains annually updated positional information for celestial objects.

In the twenty-first century, we know with exquisite detail the temperature structure of the Sun (nearly 20 million degrees Fahrenheit at the core; about 10,000 degrees Fahrenheit at the surface) and the contents of the Sun (almost entirely hydrogen and helium; almost entirely plasma—atoms stripped of one or more electrons). We have several satellites that continually measure and monitor solar activity (e.g., flares, sunspots, coronal mass ejections). The Sun has nowhere to hide life of any kind, and the intense ultraviolet and X-ray radiation fields produced by the Sun would quickly destroy any life that came near it. Without any doubt, the Sun, like the Moon, is void of life of any kind.

why mars?

With the two brightest objects in the heavens eliminated as possible locations for life, the obvious next places for astronomers in the post-Galileo era to examine as possible abodes for life were the other planets—Mercury, Venus, Mars, Jupiter, and Saturn. (Uranus and Neptune were not discovered until 1781 and 1846, respectively; Pluto, if you still include it in the family of planets, was not discovered until 1930.)

What about Mercury? As seen through telescopes in the seventeenth century, Mercury was small and featureless. In fact, as seen through telescopes through most of the *twentieth* century, Mercury was still small and featureless. Physically less than half the radius of Earth and in an orbit less than half the diameter of Earth's orbit around the Sun, Mercury is never closer to Earth than about 60 million miles (about 250 times farther away than the Moon, when the Moon is closest to Earth). The combination of Mercury's small physical size (just a bit less than twice the diameter of the Moon) and significant distance from our telescopes means that it always appears more than one hundred times smaller in angular size than the full Moon. In addition, as an inner planet, Mercury goes through phases. When Mercury is closest to Earth, and therefore should appear largest, we can only gaze at darkness, at the unilluminated side of the planet. Half an orbit of Mercury later, when Mercury should be at

"full" phase, our view of the planet is blocked by the Sun. In between, when we can actually observe Mercury, we see only a crescent.

We simply cannot learn much, if anything, from an object that
always appears so little and is merely a sliver, rather than a full disk,
when we can observe it. Also, because of Mercury's small orbit
around the Sun, as seen from Earth, Mercury never strays more than
28° (about 35 million miles) away from the Sun. Sometimes Mercury rises just before the Sun, during morning twilight, and appears
in our daytime sky just above our eastern horizon; at these times, the
sky quickly becomes too bright for us to be able to see the planet
well. At other times, Mercury sets just after the Sun, during evening
twilight; thus, just as the bright sky gives way to darkness, making
seeing Mercury possible, Mercury scurries after the Sun and disappears below the horizon. Because Mercury was assumed to be hot,
since it was near the Sun, and was small and hard to observe, eigh-
teenth- and nineteenth-century astronomers never invested much
passion into speculations about life on Mercury.

What we do know about Mercury today confirms that this planet
is inhospitable to life. At such a close distance to the Sun, Mercury is
toasted by sunlight, to a temperature as hot as 801°F (427°C) at the
point on Mercury's surface where the Sun is directly overhead. Mercury also spins very slowly, with a Mercurian day lasting nearly 176
Earth days.* With no atmosphere to blanket the surface and smooth
out day-to-night temperature differences across longitudes and latitudes, the side of Mercury that faces the Sun for months at a time is
baked to a crisp. Meanwhile the other side of the planet is chilled to
the temperatures of the depths of interstellar space, with temperatures as low as –279°F (–173°C). Thus, the surface of Mercury is likely
either too hot or too cold for life. It offers no in-between.

The small size of Mercury combined with a location so close to
the Sun has another negative effect on the ability of this planet to be
hospitable to life. Small size means much less mass (about 5.6 percent) than that of Earth. Mercury's small mass in combination with
its size yields a gravitational strength at its surface of 38 percent that

---

*The Mercury "day," which is 175.97 Earth days in length, results from the curious combination
of Mercury's slow rotation period (58.65 days) combined with its orbital period of 87.97 days.

of Earth's gravitational pull, which is nearly identical to that of Mars. As a result of their relatively weak gravitational pulls at their surfaces, holding on to atmospheres is difficult for both Mercury and Mars. The high daytime surface temperature of Mercury, however, means that atmospheric atoms or molecules would have much higher velocities in an atmosphere of Mercury than they would have in the much colder atmosphere of Mars. Higher velocities, in turn, mean that the contents of a Mercurian atmosphere would rise up and away from the surface very easily and then escape entirely from the planet into interplanetary space. In addition, fast-moving particles that travel outward from the Sun at speeds of up to a few hundred miles per second (particles entrained in what is known as the solar wind) help accelerate atmospheric particles to escape velocity from Mercury. Consequently, Mercury is unable to hold on to an atmosphere or retain water on its surface. (With one interesting exception: modern observations indicate that Mercury has a very small amount of water ice, permanently hidden from sunlight, in deep impact crater basins near its north and south poles.) Mercury is left with only an exosphere, this being a tenuous atmosphere of light gases that have been kicked off the surface and are transitioning toward escape.

And Venus? Venus is the closest planet to Earth. It is also within 5 percent of being the same physical size (radius) as Earth and has 81 percent of the mass of Earth. Yet despite its proximity, similar size, and similar surface gravity (90 percent) to Earth, astronomers mostly ignored Earth's nearest planetary neighbor because, through their telescopes, it showed nothing worthy of study.

Before the modern era of planetary exploration, Venus was of interest to astronomers for only a little while, and almost never for reasons associated with speculations about extraterrestrial life. As Galileo discovered within months of pointing his first telescope toward the heavens in 1609, Venus changes in apparent size and shape when viewed through a telescope. Venus appears bigger when it gets closer to Earth and smaller when it recedes to distances farther from Earth. As Galileo also discovered, Venus goes through crescent and gibbous phases, sort of like the Moon, except that Venus appears much bigger when in crescent phase than during gibbous phase, whereas the Moon is nearly the same size during all phases.

These changes were of revolutionary importance. Galileo explained that the way in which Venus changes in size synchronously with its peculiar changes in phase proved that Venus orbits the Sun, not Earth. He was right. Galileo's observations of Venus demonstrated unequivocally that one of Aristotle's fundamental ideas about the structure of the universe, that Earth was the center of *all* orbits, was wrong. Therefore, Galileo asserted, in agreement with Copernicus, that Earth and all the other planets also orbit the Sun. While Galileo was right about these ideas, to his eventual consternation the pattern of the changing phases of Venus implied but did not prove that Copernicus was right. Galileo argued eloquently in support of the idea that Holy Scripture should be re-interpreted and understood in the context of a Sun-centered rather than an Earth-centered universe. But the sum total of Galileo's observations and logic was insufficient to convince the leaders of the post–Reformation-era Roman Catholic Church to abandon Aristotle's geocentric model of the universe, which they believed was both logically and theologically proven correct by virtue of a long tradition of Church leaders having interpreted several lines of Scripture (e.g., "The sun rises and the sun sets, and hurries back to where it rises"; Ecclesiastes 1:5) in that sense. As a consequence of Galileo's attempts to change the Church's views on what he believed was an astronomical matter but which the Church believed was a theological issue, the Congregation of the Index in 1616 prohibited and condemned all books that claimed that Copernicus was right (ironically, Copernicus's own book, *On the Revolutions of the Heavenly Spheres*, was merely "suspended until corrected"; corrections were issued in 1620). Seventeen years later, in 1633, the Congregation of the Holy Office (i.e., the Roman Inquisition) put Galileo on trial, convicted him of heresy, placed him under house arrest for the remaining eight years of his life, and created an atmosphere of tension between science and religion with which we still live.

Yes, Venus was important to the history of astronomy and science, and those few observations of Venus made by Galileo are among the most important and revolutionary discoveries ever made in the history of science. They do not, however, turn Venus into an interesting object for further study or render it interesting for speculative thoughts on extraterrestrial life. Other than changing shape from a large crescent to a small but nearly round Venus and then back again to a large crescent, Venus appears always identically gray and featureless. These changes in size and shape of Venus mean that when it is close and large, we only see a sliver of the planet, and when Venus is waxing it is also becoming more distant from Earth and therefore harder to study. As a result, no matter how big Earth-bound astronomers made their telescopes in order to enlarge their views of Venus, and no matter how carefully they engineered their imaging filters, even into modern times astronomers sitting comfortably on Earth have never been able to see a crater, mountain, or anything else on the surface of Venus. They also have never even seen a dark or bright spot or any other kind of spot on the visible disk of Venus using their telescopes on Earth. For astronomers using telescopes over the last four hundred years, Venus has been boring and unknowable and never became the focus of our astronomical desire and attention in the way that Mars did. In fact, until the invention of ultraviolet filters for telescopes in the early twentieth century, astronomers were unable to see any features of any kind even in the atmosphere of Venus. After the phases of Venus were understood, astronomers lost interest in that planet for more than three hundred years.

Science fiction writers found Venus a bit more interesting, perhaps in large part because the thick atmosphere precluded astronomers from proving them wrong about what might be taking place on the surface. Edgar Rice Burroughs wrote his *Carson Napier of Venus* series of books, beginning with *Pirates of Venus* (1934) and continuing with *Lost on Venus* (1935) and *Carson of Venus* (1939), all set on a tropical Venus; H. P. Lovecraft's *In the Walls of Eryx* (1939) is set on a muddy Venus; and Ray Bradbury's *All Summer in a Day* (1954) is set on a rainy Venus. Twentieth-century astronomy, however, slowly but surely showed that Venus is uninhabitable, even for science fictional creatures.

In 1942, before modern astronomers had figured out how to actually measure the surface temperature of Venus, the eminent British astrophysicist James Jeans described Venus as "so hot that water would boil away . . . so that the planet may well be devoid of water."[15] He couldn't prove that assertion, but two decades later Carl Sagan would prove him right, using radio telescope measurements to determine that the daytime surface temperature on Venus was about 890°F (the best determined answer, today, is about 870°F, or 465°C).[16] Jeans also suggested that there were good reasons to think that the clouds might be made of formaldehyde. He would later be proven sort of correct, in that the clouds are made of caustic stuff, though they are not made of formaldehyde. Being open-minded, Jeans didn't completely dismiss the possibility of life on Venus, though he suggested, "any life that this planet may harbor must be very different from that of the Earth."

In the 1970s, astronomers finally discovered the reason for Venus's reluctance to reveal its surface to us. In addition to having an atmosphere that is ninety-five times denser than Earth's atmosphere and that is composed almost entirely of carbon dioxide, Venus is shrouded by a 20-kilometer thick layer of clouds. Using telescopes on Earth to observe Venus, astronomers were able to determine that the cloud droplets are composed mostly of sulfuric acid, which is nasty stuff for sure, though not formaldehyde. As a result of these thick clouds, in visible light we can never see through the atmosphere to the surface of Venus. The light that we see with our eyes, and that seventeenth-, eighteenth-, nineteenth-, and twentieth-century astronomers would have seen and twenty-first-century astronomers still see with their telescopes, reflects off the Venusian cloud tops, which are blandly yellow and almost completely featureless. In order to see through the Venus atmosphere, NASA engineers, when building the Pioneer Venus orbiters, which orbited Venus in the late 1970s, and the Magellan orbiter, which visited Venus in 1989, put cloud-penetrating radar on these spacecraft.

Using data from Pioneer Venus, planetary scientists proved James Jeans correct. Venus lost almost all of the water that it once possessed billions of years ago. Venus once had oceans and rivers. Now it is dry as a bone, with at most enough water vapor in its atmosphere to form a global ocean (if all of it were able to condense onto the surface) with a depth of only an inch or two (as compared to Earth's oceans, which are on average about 2 miles in depth).[17] And though the surface of Venus, which has craters, mountains, and continents, shares similarities with the surface of Earth, Venus is as hostile to life as any planet could possibly be.

Et tu Jupiter? Jupiter was different. Galileo had discovered Jupiter's four large moons in 1610, but through his telescope the planet itself showed nothing that drew his attention. And as seen through his and others' seventeenth-century telescopes, Io, Europa, Ganymede, and Callisto were all nothing more than white, starlike dots dancing continuously around Jupiter. Once the orbital periods of these moons were pinned down to a reasonable accuracy, little more could be learned about them until modern times, when NASA sent the Pioneer 10 and 11 and then the Voyager 1 and 2 spacecraft hurtling past Jupiter in the 1970s. We have continued to learn more about Jupiter via the Galileo orbiter in the 1990s and the currently active Juno mission.

The dark belts and zones on Jupiter were observed possibly as early as 1630 by the Italian Francesco Fontana. Three decades later, in about 1664 and almost half a century after Galileo's discoveries of the Galilean moons, English scientist Robert Hooke claimed to have seen a spot on Jupiter. A year later Cassini first described a "permanent spot" of extraordinary size on Jupiter. But Jupiter is so far away from Earth that until the advent of modern telescopes, these features were not seen in rich enough detail to generate the kind of interest that would lead astronomers to become as fascinated with Jupiter as happened with Mars.

For a few decades in the early 1600s, Jupiter was a great source for speculation about extraterrestrial life, with none other than Johannes Kepler, who invented modern mathematical astrophysics when he discovered that planets orbit the Sun in ellipses (not circles) with the Sun at one focus of those ellipses. Kepler suggested, in response to Galileo's discovery of Jupiter's four large moons, that the mere existence of the Jovian moons provides proof that Jupiter itself is inhabited.[18] Why else, Kepler explained, would God have created those moons, if not for the enjoyment of Jupiterians?

Jupiter, however, is quite far from the Sun. Consequently, the amount of heat received from the Sun by any Jovian life-forms in the upper layers of Jupiter's atmosphere would be far below that needed to maintain water as a liquid. Below the cloud tops of Jupiter, however, temperatures rise with depth in the atmosphere, such that the temperatures are above the freezing point of water at a depth of a few hundred miles. At those depths, however, the pressure is already nearly twenty times greater than the surface pressure of Earth, and pressures and temperatures continue to rise dramatically at greater depths. Could life survive and prosper in the upper atmosphere without liquid water? Could life in the deep atmosphere endure the tremendous pressure generated by Jupiter's enormous mass and gravitational strength? At best, Jupiter is an incredibly tough environment for life. Consequently, with few exceptions astronomers invested very little time and energy into speculating about Jovian life-forms.

Though prior to the 1970s, Jupiter's large moons were of no interest to astronomers interested in exploring ideas about life in the universe, in recent decades Jupiter's moon Europa has generated tremendous curiosity from planetary scientists interested in the possibility of extraterrestrial life. Soon after Europa formed, the internal heat generated by the collisional processes that formed the moon became sufficient to soften Europa on the inside. As a result, light materials (water) inside this moon rose toward its surface and the denser materials (metals, rock) descended toward the center in a process planetary scientists call *differentiation*. All of Europa's water migrated upward through the internal slush. The surface layer that formed, and that is exposed to the chilling temperatures of space, froze and formed a thick layer of ice. At the same time, Europa's rock settled downward to form a rocky core.

After the formation epoch of Jupiter and its moons, the tides generated on its innermost large moons by the gravitational interactions between these moons and Jupiter began to push Io, Europa, and Ganymede outward, synchronously. Jupiter squeezes and pushes the innermost large moon Io, Io pushes Europa, and Europa pushes Ganymede. As Io orbits Jupiter, the tidal interaction between Jupiter and Io squeezes and unsqueezes Io, as one might squeeze and unsqueeze a tennis ball. This continuous process pumps energy into and warms the interior of Io, which has become so hot that volcanic activity occurs continuously on its surface. Like Europa, Io differentiated; however, any water that long ago rose to the surface of Io has long since been heated and blasted off into space.

Io, however, transfers enough tidal heating to Europa to soften Europa's surface ices—the absence of identifiable impact craters is testament to the malleability of Europa's surface. Below that soft surface layer, which extends to a depth of a few to a few tens of miles, the heat has completely melted the ice. As a result, in between the rocky core and the thin surface veneer of ice, Europa has a global ocean. This global ocean, which might be as thick as 60 miles, is a subsurface region where the temperature, pressure, energy, and salinity conditions may be just right for sustaining life. For these reasons, Europa is great interest to modern astronomers; these internal conditions of Europa, however, were unknown prior to the modern era of space exploration, and so Europa was of no interest to premodern astronomers contemplating life in the universe beyond the confines of Earth.

If Jupiter is a harsh home for ET, Saturn is worse. Twice as far from the Sun as Jupiter, Saturn is much colder than Jupiter but is comparable in gravitational harshness. Saturn drew Galileo's interest right away, when in July of 1610 he discovered, as he described the phenomenon, that Saturn was a composite of three connected objects. He thought he had discovered that Saturn, like Jupiter, had moons, one on each side. But while Saturn's bulges appeared and disappeared, they were not small spots of light separated by constantly changing short expanses of dark sky from the planet, as were Jupiter's four large moons. Instead, they were more like handles that remained permanently attached to Saturn. Galileo could never quite figure out what he had observed at Saturn. Half a century later, Huygens, who had already discovered Saturn's largest moon, Titan, correctly deduced that a thin, flat ring surrounds Saturn. Having one huge moon and a ring system made Saturn interesting, but Saturn itself, like Mercury, Venus and, for the most part, Jupiter, showed almost no surface or atmospheric features sufficient to draw and retain the long-term attention and interest of astronomers. Since giant Saturn, like massive Jupiter, would be a cruel mother, astronomers long ago stopped expecting to find life on Saturn.

Saturn's largest moon Titan is similarly so far away that it was impossible to study in much detail from Earth prior to the advent of modern interplanetary spacecraft. These robotic explorers, most importantly the Huygens probe designed by the European Space Agency, which was carried to the Saturn system by NASA's Cassini spacecraft and dropped into Titan's atmosphere in late 2004, and the Cassini mission itself, which studied Saturn and Saturn's rings and moons for 13 years and whose mission ended with a fatal dive into Saturn's atmosphere in September 2017, have allowed us to pull back the curtain on Titan and discover the secrets of its atmosphere and surface. Modern astrobiologists speculate about the possibility of exotic life-forms living in the liquid ethane and methane rivers, lakes, and seas that dot the surface of Titan or that might have once lived in Titan's super-salty subsurface ocean of water.[19] But as with Europa, our infatuation regarding the potential habitability of Titan is a modern phenomenon.

Similarly, because the Cassini spacecraft allowed planetary scientists to study Saturn's moon Enceladus with regularity for more than a decade, Enceladus has emerged as a magnet for the attention of astrobiologists. Like Europa, Enceladus has a global subsurface ocean beneath its ice shell.[20] Enceladus also has Yellowstone-like geysers that vent water vapor, hydrogen, nitrogen, methane, and carbon dioxide into space from pressurized subsurface chambers. Some of these eruptions extend more than 300 miles above the moon's surface. As with Europa and Titan, Enceladus has grabbed the attention of modern astrobiologists, but until the beginning of the twenty-first century, Enceladus was simply a tiny, obscure moon in the outer solar system.

We did not even know that Uranus existed until Herschel, during a telescopic search for faint stars in the constellation Gemini, accidentally discovered it in 1781. In contrast, Neptune was discovered in a planned search of a precise location of the sky, conducted at the Berlin Observatory by Johann Gottfried Galle in 1846, after two mathematicians, first John Couch Adams of England and then Urbain Le Verrier of France, used the orbital positions of Uranus as measured since its discovery to infer that a heretofore unknown and more distant planet was disturbing the orbit of Uranus.[21] These two giant planets are so far from the Sun that we knew almost nothing about them until Voyager 2 flew past them both in the late 1980s. Merely small dots of light in nineteenth-century telescopes, both Uranus and Neptune entered astronomers' consciousness at a time when the idea of life on the two previously known giant planets,

Jupiter and Saturn, had dwindled. As two more giant planets, neither Uranus nor Neptune has ever generated any interest among astronomers as places where life might exist.

In addition, the nineteenth century ushered in an epoch of intellectualism during which the concept of pluralism—shorthand for the idea that large numbers of populated worlds exist—was called into question. According to this hypothesis, which had been in vogue among astronomers and philosophers for several centuries, every world in the universe—a world being synonymous with a star, planet, or moon—was necessarily inhabited because God would not have wasted creative energy in making these worlds if not to place thereon inhabitants who would worship Him.

David Rittenhouse, of Philadelphia, was one of the first to suggest that Christianity and pluralism were incompatible when he delivered his *Oration* to the American Philosophical Society in 1775. Rittenhouse suggested that "The doctrine of a plurality of worlds," which meant, to astronomers, that all distant worlds were populated by intelligent beings, "is inseparable from the principles of astronomy, but this doctrine is still thought . . . to militate against the truths asserted by the Christian religion."[22] Rittenhouse was suggesting that one could not be a Christian and also believe in extraterrestrial life. Thomas Paine spread this belief broadly when he wrote, in his *Age of Reason* in 1793, "to believe that God created a plurality of worlds at least as numerous as what we call stars, renders the Christian system of faith at once little and ridiculous and scatters it in the mind like feathers in the air."[23] In 1853, William Whewell, the master of Trinity College at Cambridge University by appointment from British prime minister Robert Peel (he later served as vice-chancellor of the University), published *On the Plurality of Worlds: An Essay*, in which he argued quite forcefully that the scientific foundations for the pluralist position "were scientifically defective and religiously dangerous."[24] Whewell was one of the most influential public intellectuals in nineteenth-century England. He invented the word "scientist" to replace "natural philosopher," served as president of both the British Association for the Advancement of Science and the Geological Society, and was named a fellow of the Royal Society.[25] Whewell's enormous influence in the academy generated a strong response, and though it did not dissuade most astronomers from embracing pluralism, it forced many to think more carefully about the question of extraterrestrial life.

Perhaps (mis)led by their biases, from the earliest days of peering through telescopes astronomers found or invented reasons to think Mars was similar to Earth. The more they studied Mars, the more convinced they became of the reality and importance of those similarities. According to their logic, since Earth has life and since Mars is similar to Earth, Mars is likely to harbor life. Furthermore, Mars, unlike the Moon, the Sun, Mercury, Venus, Jupiter, Saturn, Uranus, and Neptune, was immune to the anti-pluralist arguments of Rittenhouse, Paine, and Whewell. While in the early twenty-first century, Jupiter's moon Europa and Saturn's moon Enceladus may have surpassed Mars as possibly the best candidate bodies in our solar system for harboring extraterrestrial life, in the early nineteenth century and continuing for most of the twentieth century, Mars was the only object in our solar system for which some serious scholars, albeit amid contentious debate, continued to assert that extraterrestrial life did and might continue to exist in our solar system.

3

---

mars and earth as twins

Astronomers armed with optical telescopes fell in love with Mars during the very earliest years of the age of telescopes. Mars was bright, changed colors, and had intriguing, contrasting brighter and darker regions. Furthermore, because Mars is closer to Earth than all other planets except Venus, it could appear quite large when viewed in a telescope. In addition, unlike both nearby Venus and Mercury, which sometimes appear as mere slivers, Mars is always fully illuminated (a "full" Mars). In comparison to the other known planets in the solar system, Mars lured astronomers in and kept their attention. By the end of the eighteenth century, knowledge gained by astronomers about the physical properties of Mars would convince them that because the two planets shared so many properties, we could think of them as twins.

Almost as soon as telescopes became common, first in Italy and then elsewhere in Europe, observations of Mars took center stage. Amazingly, what astronomers saw on Mars resembled what they thought they would see were they to view Earth from a distance through a telescope. One of the most important early observations of Mars is the first claimed telescopic detection of a spot on its surface. In 1636 and again in 1638, a Neapolitan lawyer, optician, and amateur astronomer, Francesco Fontana, discovered a dark spot, a "very black pill" he called it, almost exactly in the middle of the disk of Mars. Because Fontana also observed a similar "pill" on the disk of Venus in observations made a decade later, his little "black pill" on Mars was almost certainly an optical illusion, created by the poor optics of his homemade telescope, though neither he nor his contemporaries understood that at the time.[1] Fontana's dark spot is important, not because it was a Martian spot, since it was not, and not because it was specious, but because it drew the attention of other astronomers to Mars as an object of interest with the potential to show to careful observers something other than a bland, featureless disk. The message seemed clear: Mars had secrets to share, and astronomers armed with telescopes could peer through the depths of space, pull back the Martian veil, and unravel the mysteries of the red planet.

The next report of Martian spots was of likely actual surface features on Mars. Father Daniello Bartoli, a Jesuit observing with his telescope on Christmas Eve, 1644, and like Fontana observing from Naples, detected two patches on the "lower part" of Mars. He recorded no other observations of Mars, but did preserve for us one other personal observation. "God willing," he wrote, "future observers might be able to see them better."[2] The Jesuit preoccupation with Martian spots transitioned from Naples to Rome in the ensuing decade. There, Father Giambattista Riccioli and Father Francesco Grimaldi, working together at the Collegio Romano in Rome, reported that they observed patches on Mars on multiple nights in 1651, 1653, 1655, and 1657.[3]

Mars is best positioned for viewing by observers on Earth about every two years, when Earth catches up to the slower moving Mars and both are on the same side of the Sun and aligned in the same direction, as seen from the Sun. This two-year cycle is why Riccioli and Grimaldi had recorded observations of Mars every other year through the decade of the 1650s. This alignment of both planets on the same side of the Sun, which in the language of astronomy is known as *opposition*, occurs for Earth and Mars every 780 days, and is simply the result of Earth orbiting the Sun faster (365.25 days) than Mars (686.98 days). Like two runners on a track, with Earth on a smaller, inner lane, Earth catches up to Mars every two years plus fifty days as they orbit the Sun at different angular speeds. At these moments of opposition, Mars is closer to Earth than at other times and therefore looks bigger to terrestrial-bound observers; in addition, due to the

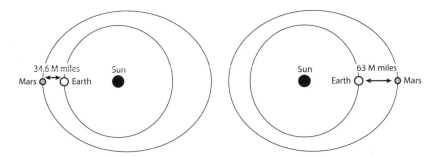

Figure 3.1. When the Sun, Earth, and Mars are aligned (Mars is "at opposition," which occurs every 780 days), the distance from Earth to Mars can be as small as 34.6 million miles (left panel) or as large as 63 million miles (right panel).

relative positions of the Sun, Earth, and Mars at opposition, sunlight reflected off Mars and back to Earth is reflected more efficiently at opposition than at other Sun-Earth-Mars orientations. For observers peering through telescopes, a bigger and brighter Mars is a better Mars because they can identify finer details on the Martian surface. At its closest, Mars can be as close as about 35 million miles from Earth; however, because the orbit of Mars is more elliptical (less circular) than the nearly circular orbit of Earth, the closest approach of Mars to Earth is not always the same distance. When at opposition, the two planets can be as close as 34.6 million miles (with Mars having an angular size of about 26 seconds of arc) but as far apart as 63 million miles (with Mars having an angular size of only about 13 seconds of arc). The time between close oppositions, which are the best times for making astronomical observations of Mars, is 15–17 years.

In the late 1650s, because of the confluence of the fortuitously good positioning of Mars for observations and the arrival on the scene of one of the most brilliant astronomers of the century, Christiaan Huygens of the Netherlands, a Martian revolution was about to occur. Mars had been at a very close opposition in 1655, and it was still fairly well positioned four years later, in 1659, when Huygens made a discovery about Mars that was formative in molding astronomers' perceptions of Mars as a twin of Earth. During the early evenings of November 28 and December 1, 1659, Huygens drew sketches of Mars that show a large, broad, dark "V"-shaped area that

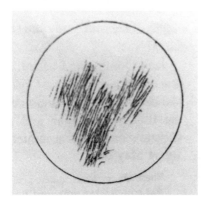

Figure 3.2. Sketch of Mars made by Christiaan Huygens on November 28, 1659. The dark patch in this sketch, first seen by Huygens and one of the most easily identified features on the surface of Mars, is now known as Syrtis Major Planum. In the nineteenth century, it was variously known as the Hourglass Sea (or Mer du Sablier), the Atlantic Canale, and the Kaiser Sea. Image from Flammarion, *La Planète Mars*, 1892.

covered about half the breadth and half the height of the visible disk of Mars. Based on the difference in times between his observations on the two evenings and the slight shift in position of the dark patch from the night of November 28 to the night of December 1, Huygens drew a bold and correct conclusion: the rotation of Mars, like that of Earth, has a period of about 24 hours.[4]

Let that discovery marinate a moment. Mars rotates, Mars has a day and night, and the length of the cycle of day and night on Mars is nearly identical to the 24-hour cycle of day and night on Earth.

The rotation period of Mars did not have to be 24 hours. After all, the rotation periods of planets in our solar system range from a few hours to hundreds of days. Jupiter spins in 9.9 hours. A "day" on Neptune is 16.1 hours. The rotation period of Pluto is 6.4 days, and Venus requires 243 Earth days to complete one turn on its axis. (None of these rotation periods were known to astronomers in 1659.) Why 24 hours for Mars? For astronomers in the seventeenth century, the answer was obvious: *Mars is Earthlike*. With this discovery, the appeal of Mars as *the* important planet for study mushroomed.

In 1666, Italian and papal astronomer Giovanni Domenico Cassini was in the process of reestablishing himself in Paris as Jean Dominique Cassini, the first director of the Paris Observatory, appointed by King Louis XIV. The Paris Observatory, however, was not yet complete. As a result, in February and March Cassini remained in Bologna, where he carried out a series of observations of Mars. Not only did Cassini see two dark patches on Mars, he observed them as they moved from east

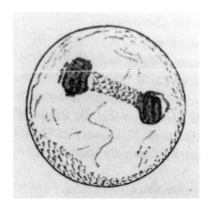

Figure 3.3. Sketch of Mars made by Giovanni Cassini in early 1666. Image from Flammarion, *La Planète Mars*, 1892.

to west across the observable disk of the planet, day after day. Curiously, these dark patches did not return to their same positions exactly 24 hours later. Instead, they reached the same position 24 hours and 40 minutes later. Mars, he concluded correctly, completes one turn on its axis in 24h 40m, rather than exactly 24 hours.[5]

In 1686, French intellectual Bernard Le Bovier de Fontenelle published one of the most widely read popular astronomy books of the seventeenth century, *Entretiens sur la pluralité des mondes* (*Conversations on the Plurality of Worlds*). By 1800, *Conversations*, which describes the living beings that, at least in Fontenelle's imagination, existed on every planet, had been translated into Danish, Dutch, German, Greek, Italian, Polish, Russian, Spanish, and Swedish. It had also found its way onto the Roman Catholic Index of Prohibited Books, where it remained until 1825. Although Fontenelle did not populate Mars with intelligent Martians, he does describe a spectacular and brilliant Mars with great high rocks that store up daylight and then glow at night. In Fontenelle's imagination, these phosphorescent rocks, together with a great number of luminous birds, light up the Martian darkness. Life on Fontenelle's Mars is unlike life on his Mercury and Venus. Both his Mercurians and Venusians are scorched by the Sun; in wondrous contrast, his Martian birds live on a world of beauty: "nobody can imagine a scene pleasanter than that of rocks illuminating the landscape after sunset, and providing a magnificent light without inconvenient heat."[6] Mars, the public learned from Fontenelle, is a beautiful, pleasant, Earthlike world, richly populated with living things.

The next great advances in our knowledge of Mars and in making Mars seem even more Earthlike emerged from the observational work done by Giacomo Filippo Maraldi, the nephew of the great Cassini. Cassini hired Maraldi as an assistant astronomer at the University of Paris Observatory, where his nephew then spent his career making observations of planets. Maraldi made four major discoveries during his studies of Mars in 1704, during the closest opposition of Mars that had occurred in 15 years. Then, after patiently waiting another 15 years, he confirmed these results during the next close opposition in 1719. First, Maraldi ever so slightly improved our knowledge of the rotation period of Mars. Mars, he determined, rotates in 24h 39m rather than 24h 40m. He also established that Mars has dark patches and that, unlike the dark patches on Earth's Moon, the Martian dark patches vary in both form and location. Finally, Maraldi discovered that Mars has bright patches at the north and south poles that changed in appearance over time. In fact, the bright patch in the south, which was offset slightly from the exact southern pole, sometimes disappeared entirely, as did one bright patch in the north. Maraldi took great care to avoid trying to explain the bright polar spots, though he did conclude that the changes in appearance were due to some genuine physical change on the surface of Mars.[7] Not surprisingly, astronomers of that era did not need much imagination to surmise that these bright patches were polar ice caps, similar to those on Earth.

Further progress in our knowledge about Mars was slow through most of the eighteenth century, but exploded again with the work of William Herschel, in the 1780s. Herschel's list of important scientific accomplishments, almost all of which he accomplished with the dedicated help of his sister Caroline, is long. He discovered the planet Uranus. He proved that some stars orbit other stars in so-called binary star systems, and he used this discovery to prove that all stars are not intrinsically identical in brightness; rather, some are intrinsically faint and others intrinsically bright. This discovery seems obvious and mundane to a modern student of astronomy, but prior to 1800, other than the existence of a small number of stars that were known to vary in brightness, definitive proof that some stars were intrinsically brighter than others did not exist. Herschel also made a map of the entire sky, noting the location and apparent distance of every star he could see. Thinking that the faintest visible stars in the heavens marked the distant edge of the universe, he thought that he had mapped out the entire universe. Herschel's universe turns out to be a map of part of our galaxy, the Milky Way, not of the entire universe, though astronomers would not understand that until the 1920s, when Edwin Hubble revolutionized astronomy with his work. Herschel also discovered the existence of light outside of the range at which our eyes can see; he found light beyond the red that we now call infrared light. He did this by passing sunlight through a prism and measuring the amount of heat absorbed by thermometers illuminated exclusively by blue light, yellow light, and red light, respectively. He then placed a fourth thermometer just beyond the visible red part of the spectrum where it apparently was not exposed to any direct sunlight; he found that this thermometer also absorbed heat from the Sun, which he correctly concluded arrives at Earth in the form of a color of light to which our eyes are not sensitive. That extensive but not exhaustive list of important discoveries and measurements justifies his historical position as the greatest astronomer of the eighteenth century.

With regard to Mars, Herschel made a number of very careful observations, one of which took astronomers of his age one step farther down the road in turning Mars into a clone of Earth. Herschel discovered that the sizes of the bright northern and southern polar spots, the presumed ice caps first seen by Maraldi a century earlier, waxed and waned anti-synchronously. When the northern polar spot shrinks, the southern polar spot grows. When the northern polar spot expands, the southern polar spot contracts. That pattern of behavior, he proposed, is due to *seasonal* changes in the ice caps. If Herschel could prove this idea correct, then he would be able to demonstrate that Mars not only has seasons, but also has northern and southern hemispheres that have seasons that occur exactly a half (Martian) year apart, just as do the seasons in the northern and southern hemispheres on Earth.

From his careful and extensive observations made across the years from 1777 through 1783, Herschel was able to prove that the rotation axis of Mars is tilted relative to the plane of its orbit around the Sun (astronomers call this property of a planet the *obliquity*) by 28.7°, and he measured the rotation period to be 24h 39m 21.67s. He was just a bit off on both measured values. The obliquity is closer to about 25.2° and the rotation period is closer to 24h 39m 35s. But we should give him credit for work well done. More importantly, the 25° obliquity of Mars is nearly identical to the 23.5° tilt of Earth's rotation axis relative to Earth's orbital plane around the Sun. This tilt for Earth, and not the changing distance of Earth to the Sun, is the primary reason for the existence of seasonal changes on Earth. Thus, Mars, with an obliquity nearly identical to that of Earth, also must have spring, summer, fall, and winter, and those seasons will occur at opposite times in the northern and southern Martian hemispheres, just as they do on Earth, where summertime in Australia occurs during wintertime in Alaska. Herschel's discovery of the obliquity of Mars proved, almost certainly, that Mars's bright polar spots were ice caps. (Exactly what kind of ice would be debated for another century.)

Mars was becoming more and more Earthlike with every new discovery. Herschel concluded, in his Second Memoir, read to the Bath Philosophical Society in England on March 11, 1784:

> The analogy between Mars and Earth is certainly more evident than for any other planets in the Solar System. Their diurnal movement [the length of the day] is almost the same; the obliquity of the ecliptic, causing the seasons, is analogous; of all the superior planets [those further from the Sun than Earth], the distance of Mars from the Sun most nearly resembles that of Earth, and as a result, the length of the Martian year is not enormously different from ours.[8]

Finally, Herschel concluded that Mars has an atmosphere. On the one hand, because of the variations in brightness in the planet in certain regions that he attributed to the presence of clouds and vapors in the atmosphere, he concluded that the atmosphere was considerable. On the other hand, he could observe stars that appeared as close as 3 to 4 minutes of arc (between 1/20th and 1/15th the angular diameter of the full moon) to the limb of the planet that did not change at all in brightness as they moved closer to Mars. From these

Figure 3.4. Hubble Space Telescope image of Mars, as seen in 2001 at a distance of 43 million miles (68 million kilometers) from Earth. Ice is evident at the southern polar cap (bottom) while a dust storm obscures the northern polar cap (top). A second giant dust storm can be seen in the Hellas Basin (lower right). Water ice clouds are seen surrounding the north polar cap, extending northward from the south polar cap, and near the Martian equator. See Plate 2. Image courtesy of NASA and the Hubble Heritage Team (STScI/AURA).

observations, he concluded that the Martian atmosphere did not extend very far from the surface; otherwise the Martian atmosphere would have blurred and extinguished the light of these stars when Mars passed so near to them.

The last major contributions to our knowledge of Mars that occurred before a group of nineteenth-century astronomers would begin mapping the surface of Mars were made by the accomplished German astronomer Johann Hieronymus Schröter, who had his own observatory in the city of Lilienthal, where he was the chief magistrate. Schröter's list of astronomical accomplishments is lengthy. He was the first to prove that Venus has an atmosphere, and he was one of six astronomers, self-named the Lilienthal Detectives, or the Celestial Police, who set out together to discover the supposedly missing planet orbiting the Sun in between the orbits of Mars and Jupiter. Ultimately, over a period of only seven years, from 1801 through 1807, members of this distinguished band of astronomers discovered four such objects—Ceres, Pallas, Juno, and Vesta—in the part of the solar system now known as the asteroid belt.

Schröter observed Mars almost continuously for 18 years, from 1785 through 1803, producing 230 different drawings of it. He confirmed most of Herschel's discoveries, getting similar but slightly different values for the obliquity (27.95°) and the rotation period (24h 39m 50s). Schröter's most important new contributions to our knowledge of Mars were his observations of continual changes, sometimes hourly, in the patterns of the dark patches on Mars. These patterns were never the same, from night to night and year to year. Schröter concluded that clouds were responsible for the changing colors he observed on Mars; in fact, he came to believe that the patches themselves were entirely atmospheric phenomena rather than surface features.[9]

By the time the eighteenth century ended, two centuries of telescopic observations of Mars that had begun in the early 1600s had yielded a reasonable portrait of the red planet. Astronomers had accurately measured the rotation period, the axial tilt, the existence of polar caps that waxed and waned with the seasons, and the presence of a thin atmosphere with clouds that could, at times, obscure parts of the surface. What they concluded was self-evident to astronomers and to anyone else paying attention: the rotation period of Mars is *just like the day/night spin of Earth*, the obliquity of Mars is *just like the tilt of Earth*, the seasonal changes of Mars are *just like the seasons we find here on Earth*, the polar caps on Mars are *just like the ice caps on Earth*, and the thin atmosphere that is sometimes transparent and at other times is opaque with clouds behaves *just like the cloudy atmosphere of Earth*. Mars and Earth, they had established, are physical twins.

4

imaginary mars

Having found so many similarities between Earth and Mars, astronomers were determined to find more. Surely, if the days, seasons, years, ice caps, and clouds on Mars are just like those on Earth, then the environment, including the contents and temperature of the Martian atmosphere, all of which on Earth allow it to be hospitable to human life, must be similarly Earthlike.

Thus, armed with their telescopes and their desires, astronomers began imagining a Mars that was in every way like Earth: in their minds, they began terraforming Mars in the 1830s. The act of terraforming Mars would change its physical environment such that it would become an Earthlike world, with a temperate climate, running water, and a breathable atmosphere; once terraformed, humans could live on Mars. Earthbound astronomers in the nineteenth century could not, of course, actually terraform Mars; but they could quickly reshape their collective understanding of Mars, changing it from a hostile world into one where humans, butterflies, and ferns could all live. Imagination combined with herd instinct are powerful tools for self-deception. By the time they were done a half-century later, the old Mars, with nothing but a few bright and dark patches and a pair of polar caps, had been transformed completely, magnificently, into a highly evolved world with rivers, bays, seas, continents, and a planet-girdling system of canals built by an advanced and sophisticated civilization whose engineering skills far surpassed those of humans.

In their scientific imaginations, two German astronomers, Wilhelm Wolff Beer and Johann Heinrich von Mädler, carried out the pioneering work of terraforming Mars. Beer was a banker by trade, an astronomer by hobby. Mädler, his partner in astronomical work, was a professional astronomer who began his career working in a private observatory in Berlin that was built by Beer. With the benefit of Beer's ability to finance Mädler's research, Mädler had at his disposal an excellent telescope made by Joseph von Fraunhofer, one of the best opticians in the world. Fraunhofer had already proven the value of having great optical devices for astronomy, as in 1814 he had begun a study of a series of 570 dark lines in the spectrum of the Sun. These lines uniquely fingerprint the chemical composition of the Sun's atmosphere, and two centuries later professional astronomers still refer to them as Fraunhofer lines. In the 1830s, Beer and Mädler demonstrated that a telescope of very modest size—the collecting lens of their telescope was only 3.75 inches in diameter, though it had a magnification power of 185 times—but one with the ability to produce images of excellent optical quality when implemented by astronomers with patience, skill, and perseverance, could be used to reimagine an entire world.

Beer and Mädler carried out a program of repeated observations of Mars from 1831 through 1839. They then published their work in a series of papers in the premier journal of the time dedicated to the work of professional astronomy, *Astronomische Nachrichten* (founded in 1821 by German astronomer Heinrich Christian Schumacher and now the oldest astronomy journal in the world still being published). They then combined all of their work into a book they had printed, in French in 1840 and German in 1841. As part of their work, they published the first complete map of the surface of Mars, including separate maps of the northern and southern hemispheres, covering all 360° of longitude as well as all latitudes, ranging from 90° south to 90° north. In doing so, Beer and Mädler identified "a small patch of a very pronounced black . . . so well marked, and so near to the assumed equator, that we believe we should choose it as the reference point for the determination of the rotation period."[1] They then used this small dark patch, marked with the letter *a* on their maps, to define a single point on the surface of Mars as the zero meridian,

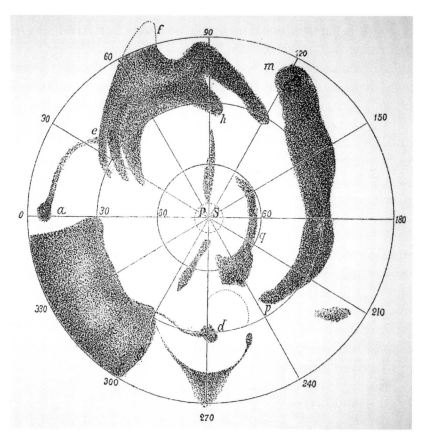

Figure 4.1. Map of the southern hemisphere of Mars, centered on the south pole, as drawn by Beer and Mädler in 1840. The feature identified as "*a*," later called Meridian Bay by Flammarion, was used by Beer and Mädler to define the origin (0 degrees) of the latitudinal system on Mars (the equivalent of the Greenwich meridian on Earth). Image from Flammarion, *La Planète Mars*, 1892.

just as on Earth we define the zero (or prime) meridian as the line of longitude that passes through Greenwich, England.

Beer and Mädler noted that though they could not make out shadows cast by mountains, given the great distance of Mars from Earth, they could identify areas on the surface that reflected sunlight differently. That is, they saw both bright and dark patches, which, they suggested, "must therefore be due to differences in reflecting power," as would also be the case "between the reflectivity of different places on the Earth." In their studies of the general "reddish colouration" of Mars, they noted how "the colour in these regions recalls that of a beautiful sunset on our Earth." This conclusion led them "with certainty toward admitting that Mars has a very appreciable atmosphere, similar to that of the Earth." The logic is straightforward (but strange): identify the red color of Mars, free-associate that red color with something red associated with Earth—sunsets; then draw the obvious conclusion that the red color on Mars is caused by sunlight passing through a thick atmosphere. The logic is flawed to the point of being ridiculous, because we now understand that red sunsets follow the setting Sun and sweep around Earth. We also know that the red color of Mars is due to the iron-rich dust on the surface. But those flaws were not evident to Beer or Mädler or those who followed up their work in subsequent decades. Beer and Mädler both intended to prove that Mars had "a very appreciable atmosphere," and they found an argument that they believed provided that proof.

They continued with a discussion of the polar patches that, they concluded without any doubt or actual evidence, are "genuinely made up of snow, shrinking with the onset of summer." As the polar ice melts and evaporates, "the surface close to the evaporating snow will become extremely humid," creating "wet, marshy soil." "It is not going too far to claim," they continued with their extreme overreach, "that Mars bears a strong resemblance to the Earth, even with regard to physical conditions." Beer and Mädler had seen on Mars, or at least they had imagined seeing on Mars, everything they could possibly have imagined they might see were they to study Earth over a period of many years through a telescope from a distance of 40 million miles.

Two decades later, at the time of the Mars opposition in 1858, Father Angelo Secchi, the director of the observatory of the Collegio Romano in Rome, decided to map Mars. On his map he identified a great blue, triangular patch he labeled the *Atlantic Canal*. In Italian, Secchi's word was *Canale*, and thus it was Father Secchi who first introduced that word into our human vocabulary for a particular kind of surface feature on Mars. He also identified another, smaller canale, which connected two broader surface patches. "These two *canali*," he wrote, "enclose a reddish continental area." As for what the Martian surface colors represent, physically, Secchi had an answer, about which he exuded extreme confidence. "The reddish regions," he wrote, "like the bluish ones, seem too permanent for their nature to be doubted; it is probable that the former are solid, the latter liquid."

Why exactly, should blue regions be liquid and red regions land? Because on Earth, astronomers of the time assumed that the bluer regions (if they could be seen from space) would be oceans and the red regions would be land. They therefore felt confident that the same color-to-material associations must hold true on Mars as on Earth. Secchi, as a result, had "proven" that Mars had Earthlike oceans and continents. He described the large, blue, triangular patch as "the Atlantic Canal, a name given for brevity to this large blue patch which seems to play the role of the Atlantic which, on Earth, separates the Old World from the New." He further discussed an isthmus, a large reddish continent, a great blue canal, and bright clouds, all of which made Mars sound very much like Earth.

Building on the work of Secchi, Emmanuel Liais, an astronomer at the Paris Observatory and later director of the Observatory of Rio de Janeiro, declared, in 1860, that the red color of Mars was due to vegetation. This declaration makes sense if, as an observer, you know that the red regions are large land areas and you also know that the land areas are covered by red-colored vegetation. Liais, of course, did not know either of these things. Perhaps the only thing he knew for certain is that on the one planet about which he had a great deal of direct knowledge, the vegetation covering the large landmasses on Earth is green, not red. That knowledge, however, did not seem to influence his certainty about what he claimed to know about Mars. On Mars, apparently, continental-sized swaths of vegetation were red.

Barely nineteen years old at the time, the young French astronomer Camille Flammarion published the first edition of his *La Pluralité des Mondes habités* in 1862. *Plurality of Inhabited Worlds*, which first appeared as a 56-page pamphlet that sold for two francs, expressed Flammarion's certainty about the existence of extraterrestrial life. The first edition sold out immediately; it also cost Flammarion his job on the staff of the Paris Observatory. The second edition, however, expanded to 468 pages, established Flammarion as the popular voice of astronomy for his time. As much as anything written in the nineteenth century, *Plurality of Inhabited Worlds*, which would go through seventeen editions just between 1864 and 1872, would color everyone's thinking about Mars and Martians, going forward. Flammarion pointed out the many similarities between Earth and Mars and then concluded that those similarities led naturally to the conclusion that Mars is inhabited by intelligent beings:

> The atmospheric envelopes which surround it and Earth; the snows which appear periodically over the poles of both planets; the clouds which extend from time to time in their atmospheres; the geographical arrangement of their surfaces in terms of continents and seas; the seasonal variations and the climates common to these two worlds; leads us to believe that both planets are inhabited by beings whose organization is of similar character.[2]

Angelo Secchi returned to his Martian studies in 1862. "Mars is the best-studied of all celestial bodies, except for the Moon," he wrote. "On it, Herschel and other astronomers have observed not only seas but continents, but also the effects of the seasons of winter and summer." Secchi went on to explain that the variations in the sizes of the polar patches and in the appearances of clouds "proved that liquid water and seas exist on Mars . . . the existence of seas and continents, and even the alternations of the seasons and the atmospheric variations, have been today conclusively proved." Secchi didn't say that he "believes" that he has found water on Mars; he didn't say he "thinks" Mars has seas and seasons. Speaking with more confidence and far less caution than a scientist of today might, "conclusively proved" were his strong, confident words, and Secchi did not hesitate to tell readers that he had "proved" many things about Mars.

Secchi then assigned an international smörgåsbord of names to certain regions that Beer and Mädler had only identified with alphabetical designations such as *e* or *f* or *h*. One he called the Cook Sea, another the Marco Polo Sea, another the Franklin Canal. A reddish area became Cabot Continent, and so the imaginative terraforming of Mars continued.

The Reverend William Rutter Dawes, both a physician and a clergyman by training, gained a strong reputation as an astronomer in mid-nineteenth-century England. His astronomical work was outstanding and influential. As early as 1850, he had discovered the so-called crepe ring of Saturn (now known as the C ring), and he is credited with making observations of the Great Red Spot on Jupiter as early as 1857, several years before the greater astronomy community recognized the existence of this Jovian atmospheric feature as fact. The Royal Astronomical Society awarded him their Gold Medal in 1855; a decade later, in 1865, they elected him a fellow of the Royal Astronomical Society. When Dawes decided to apply his observing skills to Mars, the astronomy community paid attention.

In 1865, Dawes published eight drawings of Mars, obtained during the Martian opposition in late 1864, in a contribution to the *Monthly Notices of the Royal Astronomical Society*.[3] Other astronomers were in awe of Dawes, convinced that he was able to see details beyond those accessible to other observers. After all, he had a reputation from his body of work for exactly that. Thus, the "magnificent drawings by Dawes," in the words of Camille Flammarion, "represent a considerable advance in our knowledge of Martian topography."[4]

Quite importantly, Dawes discovered that the small, round patch originally identified by Beer and Mädler was "distinctly forked . . . giving the impression of two wide mouths of a river." Dawes, however, was unable find the rivers themselves. He did conclude that "Nothing appears more certain, than that the red tint of Mars is not produced by the atmosphere of the planet; the reddish colour is always more pronounced toward the center of the disk, precisely where the atmosphere is thinnest." This concept, which implies observers are seeing all the way through the atmosphere to the surface of Mars when they see the red color, had appeared first, but was explained less clearly and effectively, in the work of French physicist and astronomer François Arago, in his *Astronomie populaire*, published posthumously over the years 1854–1857.[5] Dawes's work, however, was far more influential on the community of those studying Mars in earnest. As a result, by the late 1860s the idea that the red color of Mars represented conditions of the surface and was not an atmospheric phenomenon had become widely accepted. The bigger question that emerged, going forward, was "What is the source of the red color on the surface, plants or rocks?" Emmanuel Liais had already put forward the idea that the red color was identical with plant life. Others would follow suit.

The next step in reimagining Mars as being Earthlike in every way was taken by another Englishman, Richard Anthony Proctor, one of the great popularizers of astronomy in the 1860s and 1870s. Proctor had already secured his reputation as a professional astronomer. Like Dawes, he was elected as a member of the Royal Astronomical Society in 1866 and was an Honorary Member of King's College London. Proctor wrote books almost continuously that were widely read, including *Saturn and his System* (1865), *Planetary Orbits* (1867), *Other Worlds than Ours* (1870), *Atlas of Astronomy* (1873), and *Chart of 324,000 Stars* (1873). His obituary, published in 1888 in *The Observatory*, identified him as one "whose name as an expositor of science has become a household word wherever the English language is spoken."[6]

In 1867, Proctor published his *Charts of Mars*, in which he used the Martian drawings made by Dawes to produce a map of Mars that included "a properly worked-out system of nomenclature." That is, Proctor gave a name to every feature on the surface that he could identify. Thanks to Proctor, Mars now had four named continents—Herschel I, Dawes, Mädler, and Secchi, and two oceans—Dawes Ocean and De La Rue Ocean. The small Martian sphere had several areas he called "lands"—Cassini Land, Hind Land, Lockyer Land, Laplace Land, Fontana Land, Lagrange Land, and Campari Land, as well as several seas—Maraldi Sea, Kaiser Sea, Main Sea, Dawes Sea, Hooke Sea, Beer Sea, Tycho Sea, Airy Sea, Delembre Sea, and Phillips Sea. Dawes clearly won the popularity contest, as Proctor named a continent, an ocean, and a sea for him. Proctor identified other features as bays, forked bays, straits, islands, and ice caps. None of the names invented by Proctor remain in use. He also collected and compared all the extant measurements of the rotation period of Mars and concluded that the true rotation period of Mars* is 24h 37m 22.7s. He was right. For comparison, the modern measured value is 24h 37m 22.663s ± 0.002s.[7]

---

*This is the rotation period of Mars with respect to the stars and the entire universe, not with respect to the Sun, and is known as the *sidereal day*. A solar day on Mars (also known as a *sol*), is the length of time from one sunrise to the next, and is about 2 minutes longer than a sidereal day, or 24h 39m 35.2s. The sidereal and solar days have different lengths because Mars does not stay in exactly the same place while it spins, since it orbits the Sun as it rotates on its spin axis.

The age of imaginative terraforming Mars reached its most advanced stage with the contributions of English artist Nathaniel Green. In 1877, Green traveled to the Portuguese island of Madeira (in the Atlantic Ocean, west of Morocco) where he believed, at elevations approaching 2,300 feet and at a more southerly latitude than England such that Mars would be higher in the sky than as seen from England, that he would have excellent views of Mars. Then, during the period from the middle of August through early October, using a fairly large telescope (13-inch diameter) for that time, he set to work creating lithograph drawings of Mars as well as a composite map of the entire planet. His areographical* chart of Mars, published that year by the Royal Astronomical Society of London, was an update of Proctor's map of Mars. Green's map also had four continents, though the names had changed. Herschel, Mädler, and Secchi remained. Dawes continent, however, was gone, replaced by Beer continent. Dawes and De La Rue Oceans remained. From the viewpoint of Green, Dawes apparently couldn't have both a continent and an ocean named for him.

Now that Mars had been carefully mapped, astronomers were ready to complete their imaginative terraforming of Mars. The next group of astronomers would complete the process of creating a vision of an Earthlike Mars by finding conclusive evidence of water and plant life.

*Areography is the geography of the surface of Mars.

5

misty mars

One of the important research techniques used by mid-nineteenth-century astronomers was applying the newly invented craft of spectroscopy to the study of Mars. With the tools of spectroscopy, they discovered what they believed was proof for the presence of water on the surface and in the atmosphere of Mars. With knowledge that water exists on Mars, they believed they had proof that Mars had an Earthlike climate and that the red patches on Mars were vegetation.

Spectroscopy involves channeling a beam of light from any source through a prism or a grating (a grating can also be reflective), which spreads the light out into its constituent colors, allowing scientists to study the details of brightness and faintness of the different colors. With a simple prism, one sees a few broad colors of the visible rainbow. With a high-resolution grating, however, visible light is spread out much more broadly, into thousands (or even tens of thousands) of different shades of blue, followed by thousands of different shades of green that slowly transition into thousands of different shades of yellow, then thousands of different shades of orange, and finally, thousands of different shades of red.

In obtaining such a spectrum of Mars using a telescope on Earth's surface, we need to keep in mind that we are looking at light that originates in the photosphere of the Sun. The sunlight radiates outward through the Sun's outer atmosphere, travels through our solar system across more than 140 million miles of nearly empty space, penetrates downward through the atmosphere of Mars, reflects off the surface of Mars, passes back upward through the Martian atmosphere, then travels across another 25–50 million miles of interplanetary space before it gets to the vicinity of Earth. Finally, the light filters through Earth's atmosphere. In an actual spectrum of Mars, we will find that some of these thousands of nuanced shades of color are faint or missing because some molecule or chemical element, either in the Sun's upper atmosphere or the atmosphere of Mars or the atmosphere of Earth, has absorbed some or all of the original sunlight at exactly that very narrow shade of color. Astronomers call a region of a spectrum at which the amount of light is reduced or absent for one of these reasons an absorption line. The 570 dark lines discovered in the spectrum of the Sun by Josef von Fraunhofer in 1814 are absorption lines generated in the atmosphere of the Sun, and they provide chemical clues for us about the composition of the outer layer of the atmosphere of the Sun. With a carefully designed experiment, astronomers can deduce whether the molecule or element removing a particular shade of light in the visible light coming from Mars—all of which, remember, is reflected sunlight—is found in the atmosphere of the Sun or Mars or Earth.

The pioneering work in the spectral analysis of Mars was done by William Huggins, of the Royal Astronomical Society of London, and William Allen Miller, professor of chemistry at Kings College London. They studied Mars with a primitive spectroscope in April 1863 and again, with improved equipment, in August and November 1864. In doing so, they managed to detect several strong absorption lines at the violet (short wavelength) end of the visible spectrum that they attributed to Mars. They suggested that the red color of Mars (at the long wavelength end of the spectrum) is a consequence of Mars being effective at reflecting red light but ineffective at reflecting violet and blue light. This is the same reason red paint is red: the chemicals in red paint are good at absorbing violet, blue, green, and yellow light but are good at reflecting (or not absorbing) red light.

Huggins continued these spectroscopic studies of Mars and published additional results in 1867 as a contribution to the *Monthly Notices of the Royal Astronomical Society*.[1] By comparing spectra of Mars with spectra he obtained of the Moon, and then by identifying features in the Martian spectra that also appeared in the spectra of the reflected light from the Moon, Huggins was able to identify the spectral features due collectively to the Sun, the Moon, and Earth's atmosphere. He concluded, quite reasonably, that the spectral features that appeared in the spectrum of Mars but that did not appear in the spectrum of the Moon must be created solely by the atmosphere or surface of Mars.

Huggins detected a large number of absorption lines in the spectrum of Mars that were found in the vicinity of the Fraunhofer F line (a line in the blue part of the spectrum now known to be caused by excited hydrogen atoms). The F line, he knew, was created in the solar atmosphere. All the other lines in the Mars spectrum were absent in a solar spectrum and therefore, he knew, were due to Mars. These lines filled the Martian spectrum from the blue region all the way toward the violet end (away from the red end), and thereby removed most of the blue and violet colors from the reflected light from Mars.

Huggins now had much more information that helped him explain, more completely, what he had already come to understand in 1864. Presumably, the reason Mars looks so red, he inferred, is that most of the violet and blue sunlight that initially reaches Mars is absorbed by the Martian atmosphere in these absorption features, leaving mostly red light to be reflected by Mars. In 1864, these violet and blue absorption lines were weaker in November than in August; that is, Mars reflected more violet and blue light in November than in August. As a result, Mars was less red (because it reflected more blue and violet light) in November than it had been in August. Huggins concluded that Mars appeared redder in August when sunlight reflected off the surface and appeared bluer in November when sunlight reflected off water in the atmosphere.[2] In other words, Huggins believed that Martian mist, when present, was effective at reflecting blue sunlight; when the mist is not present in the atmosphere, sunlight reaches the surface, which is effective at absorbing blue sunlight, leaving mostly red light to reflect off the planet and back to our telescopes.

By the end of the decade of the 1860s, Huggins, using the tools of laboratory optics and chemistry applied to astronomy, had helped create a new hybrid discipline, astrophysics. Henceforth, astronomers would no longer be limited to measuring only the positions and brightnesses of celestial objects. They would learn how to use the spectra of celestial objects to discover the materials that compose the atmospheres of stars and planets; then they would learn how to use spectral signatures to determine the temperatures, pressures, densities, chemical compositions, motions, and masses of heavenly bodies. This information, in turn, combined with fundamental laws of physics, would allow them to understand the physical structures of the insides of stars, to figure out how stars are born, how they produce light, how they perform nuclear fusion to change hydrogen into heavier elements, how their internal structures evolve as they age, how long they live, and how and why and when they die. Spectroscopy, in the twentieth century, became the key to understanding the structure and evolution of the entire universe.

Huggins began the age of spectroscopy by applying these newly invented techniques of astrophysical spectroscopy to the study of planetary atmospheres, among other celestial objects. In doing so, he uncovered a remarkable result: spectroscopic evidence for the existence of water in the atmosphere of Mars. Claims for the presence of water on Mars (or in the atmosphere of Mars) no longer rested solely on the assertion that "the surface has a dark spot that looks like it should be an ocean." Now the forensic tools of physics and chemistry could be used to tease apart the light from Mars to look for spectroscopic evidence of water or other materials. Proof that water exists on Mars lent enormous credibility to the idea that Mars was Earthlike. If Mars shows spectroscopic evidence for water, then the so-called bays, seas, and oceans might, in fact, be exactly what they looked like.

The spectroscopic work of Huggins was pioneering and done well. His technique for identifying the spectroscopic features of Mars, as opposed to those of the Sun or of Earth's atmosphere, worked and still works. He did not, however, actually know what material was responsible for the plethora of blue and violet absorptions lines in his Martian spectra and thus did not have any actual evidence for the presence of water in the Martian atmosphere. That claim, though widely embraced by the professional astronomy community, was nothing more than an educated guess and, as we now know, an overinterpretation of his data. Huggins took one step too far, but others would follow his influential lead.

French astronomer Jules Janssen, who used the technique of spectroscopy to make the first known observations of helium in the atmosphere of the Sun during a total eclipse in 1868* and founded the Meudon Observatory in 1875, followed up the work of Huggins with his own imaginative, spectroscopic experiment. In 1867, he hauled his equipment to the summit of Mount Etna on the island of Sicily, at an elevation of 11,120 feet. From that location, he obtained spectra of both the Moon and Mars (and Saturn). At this high elevation, where he believed he was above most of the water in the atmosphere of Earth (he was wrong**), he hoped to minimize the effects of terrestrial water vapor in the spectra of both celestial objects. By minimizing the contamination of his spectra with the signal of terrestrial water vapor, and by comparing the high-altitude spectrum with others of Mars obtained at sea level from Palermo and of terrestrial water vapor collected at the La Villette Works in Paris, he thought he had accurately made a qualitative comparison of the water content in the Martian and terrestrial atmospheres. From his work, Janssen concluded, like Huggins, that he could identify "the presence of water vapour in the atmospheres of Mars and Saturn."[3]

Also like Huggins, William Wallace Campbell was another of the pioneers of astronomical spectroscopy. Shortly after the founding of Lick Observatory in 1888 by the University of California, the first Lick Observatory director hired Campbell as a young assistant to help senior astronomer James Keeler with spectroscopic observations. After Keeler left for Allegheny Observatory, Campbell took over as the senior spectroscopist at Lick. Campbell quickly put the

---

*Janssen, however, did no more than notice a bright yellow line in the solar spectrum. Later in 1868, Englishman Norman Lockyer noticed this same line. Helium, which is the cause of this yellow line, was first isolated in a laboratory on Earth by Scottish chemist Sir William Ramsay in 1895; Ramsay was awarded the Nobel Prize in Chemistry in 1904, "in recognition of his services in the discovery of the inert gaseous elements in air, and his determination of their place in the periodic system."

**The density of Earth's atmosphere is about 50 percent that at sea level at an altitude of 3.5 miles (18,500 feet); however, the amount of water vapor in the atmosphere varies greatly with geographic location as well as with elevation.

powerful tools at his disposal to work. One of those tools was the
Great Refractor, a 36-inch diameter telescope that achieved the goal
of its benefactor, the eccentric California millionaire James Lick,
who aspired to build a "telescope superior to and more powerful
than any telescope yet made."[4]

In 1894, noting carefully the mistakes made by Huggins and Janssen in their experimental protocols, especially in observing from
humid environments, Campbell explained that the combined factors of the dry environment of California, the largest telescope in the
world that he had at his disposal, the high altitude—4,260 feet—of
Lick Observatory, and the improved equipment he would be using,
would enable him to carry out the definitive test of whether detectable levels of water vapor were present in the Martian atmosphere.
He then laid out the criteria by which he would compare the Martian spectra with those of the Moon and how those spectra would
be obtained under identical observing conditions. After observing
Mars and the Moon on ten different nights during the months of
July and August 1894, he had his answer. "The spectra of Mars and
the Moon, observed under favorable and identical circumstances,
seem to be identical in every respect." Because the Moon was known
to not have an atmosphere, the answer was clear to Campbell. Any
absorption lines in the spectrum of the Moon were due solely to the
atmosphere of Earth. Furthermore, because the lunar and Martian
spectra looked the same, the same conclusion could be applied to
Mars. In his words, "The *atmospheric* and *aqueous vapor* bands which
were observed in both spectra appear to be produced wholly by the
elements of the Earth's atmosphere. The observations, therefore,
furnish no evidence whatever of a Martian atmosphere containing
aqueous vapor."[5] Campbell had shown, quite convincingly, that Huggins and Janssen had detected water vapor in Earth's atmosphere, not
in the atmosphere of Mars.

In November 1894, after Campbell challenged Huggins's claim
of having discovered water vapor in the atmosphere of Mars, Huggins revisited his work of three decades prior. Again, Huggins rose
to the challenge. First, he obtained photographs of spectra of both
the Moon and Mars, but he was unable to discern any differences
between the spectral features of the two objects in the November

photographs. On three nights in December, however, both Huggins and his wife made comparisons by eye of the faint spectral bands of the Moon and Mars, obtained within a few minutes of each other. "On these three nights," he wrote in an article he chose to publish in volume 1 of a brand new journal, the *Astrophysical Journal*, which was billed as "An International Review of Spectroscopy and Astronomical Physics," "the atmospheric bands ... to which our attention was almost exclusively directed, varied considerably in intensity in the Moon's spectrum, but were always estimated by us to be rather stronger in the spectrum of Mars." In repeated versions of this exercise, "Mrs. Huggins' independent observations agreed with my own." The conclusion of their work "is to leave the strong conviction in our minds that the spectroscope does show an absorption that is really due to the atmosphere of Mars." Unstated but understood by all was the idea that this absorption band* was the signature of *water vapor* in the Martian atmosphere.[6] Also unstated was the fact that despite being published in the *Astrophysical Journal*, Huggins's conclusions were based on the premodern technique of judging Martian colors using the human eye, whereas Campbell's work was modern astrophysics.

In 1908, Vesto Melvin Slipher, working on Percival Lowell's staff and on Lowell's behalf at Lowell Observatory in Flagstaff, Arizona, observed Mars from the high altitude of Flagstaff—7,250 feet. Over the course of the next few decades, Slipher would establish himself as one of the greatest observational astronomers of the twentieth century, if not of all time. Most notably, over a decade from about 1913 onward, Slipher measured the radial velocities (toward or away from Earth) of several dozen galaxies and discovered they were almost all redshifted. That is, almost all of these galaxies were speeding away from the Milky Way at velocities of hundreds to thousands of miles per second. Edwin Hubble recognized, in 1929, that Slipher's and his own more recent velocities were positively correlated with the distances of these same galaxies, meaning that more distant galaxies were

---

*An absorption band is a series of spectral lines having a common source, e.g., water molecules, in a common energy state, and that are close together in wavelength. At low resolution, these many absorption lines blend together into a single, broad absorption band.

moving away from the Milky Way, from us, more quickly than closer galaxies. Thus, Slipher's measurements of velocity redshifts of galaxies led very directly to Hubble's discovery of the expanding universe and to our understanding that the universe began with a Big Bang.

Slipher spent his entire career at Lowell Observatory in Flagstaff, Arizona. He began his work there as a staff astronomer in 1901, and, after the death of Percival Lowell, served as observatory director from 1916 until 1954. Under his leadership, Lowell Observatory hired Clyde Tombaugh in 1929. Soon thereafter, in 1930, Tombaugh discovered Pluto. Slipher developed a strong reputation for reporting his discoveries only after carefully and cautiously confirming them; eventually, according to his biographer William Graves Hoyt, he "probably made more fundamental discoveries than any other observational astronomer of the twentieth century."[7] Slipher was awarded the Lalande Prize by the French Academy of Sciences in 1919, the Henry Draper Medal by the National Academy of Sciences in 1932, and the Gold Medal by the Royal Astronomical Society in 1933.

On Mars Hill at Lowell Observatory in 1908, Slipher was observing Mars at nearly twice the elevation of his competitors at Lick Observatory, and above half of the water vapor in the atmosphere of Earth at this location. His experimental method was essentially identical to that used previously by Huggins, Janssen, and Campbell: he compared the spectrum of Mars with the spectrum of the dry, airless Moon. In his spectra, Slipher claimed to have detected a "delicate Martian component" of water. He asserted that "the reasonable conclusion is that the spectrograph has revealed the presence of water in the atmosphere of *Mars*." He then suggested that "more observations are needed before any definite statement can be made of the amount of water-vapor in the Martian atmosphere."[8] This particular research result, which formed the entirety of Slipher's PhD dissertation from Indiana University, awarded in 1909, might be the least impressive and least defensible one done by Slipher over his extended and distinguished career. Slipher never returned to or again mentioned this research result.

A year later, Slipher's professional nemesis on the water-on-Mars issue, W. W. Campbell, by then the director of Lick Observatory, led an expedition to the top of the then-highest peak in the United

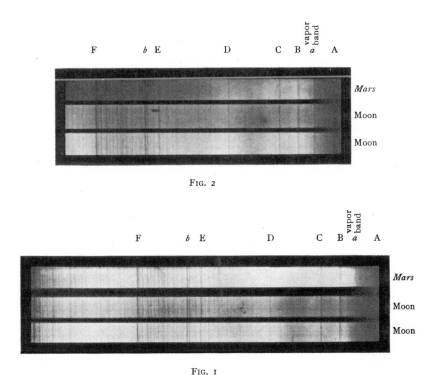

Figure 5.1. Comparison spectra of Mars and the Moon obtained by Slipher in 1908. The spectra in the bottom panel were obtained in very dry observing conditions in the air above the telescope, whereas the spectra in the top panel were obtained when the telluric moisture content was high. Slipher argued that the "vapor band" (below the letter a) was stronger (i.e., darker) in the Mars spectra than in the lunar spectra, and this was evidence for water in the Martian atmosphere. Image from Slipher, *Astrophysical Journal*, 1908.

States, Mount Whitney in southern California. There his spectro-scopic observations of Mars, made at an elevation of 14,600 feet above sea level, would be obtained at a position that would be above 80 percent of the water vapor in Earth's atmosphere at that location. Campbell found, as he had a decade before, that the so-called water vapor bands were identical in appearance for both the Moon and Mars. He concluded, cautiously and very reasonably, "This does not mean that Mars has no water vapor, but only that the quantity present, if any, must be very slight."[9] He then repeated this experiment in January and February 1910, from Lick Observatory on Mount Hamilton, at a time when the relative velocity of Mars to Earth was high enough to Doppler shift any Martian water lines away from terrestrial water lines.

The Doppler shift is a change in the wavelength of light detected by an observer because of the relative motion of the light source and the observer. If the light source (in this case Mars) is moving away from Earth, light waves from Mars are shifted to longer wavelengths (this shift of the light from yellow toward the red is called a redshift); if the light source and the observer are moving toward each other, the detected light is shifted to shorter wavelengths (a blueshift). Using this observing protocol, Campbell found that "the quantity of water vapor existing . . . in the equatorial atmosphere of Mars was certainly less than one-fifth that existing above Mount Hamilton."[10]

Half a century later, a team from the National Geographic Society and the National Bureau of Standards decided that observing techniques and equipment had improved to the point at which a definitive detection of water in the atmosphere of Mars was finally possible. C. C. Kiess, C. H. Corliss, Harriet K. Kiess, and Edith L. R. Corliss set up their equipment at a National Weather Bureau station near the summit of Mauna Loa, in Hawai'i, in 1956. Not only were they observing from an elevation comparable to that of Mount Whitney, but in addition the air above the highest peaks in the Hawai'ian islands is remarkably dry. They also took advantage of the Doppler effect, which should have moved the Martian water lines very slightly from overlapping exactly with the positions of the terrestrial water lines. The results "were negative." The "numbers of molecules [of water are] too small to produce lines of sufficient strength for micrometric and photometric measurement. . . . We must conclude that if the water vapor in the planet's atmosphere were entirely condensed, it would form a film of liquid water less than 0.08 mm thick."[11] Campbell was right. Huggins, Janssen, and Slipher were wrong. This back-and-forth, carried out over a period of a full century, captures the essence of science at its best: scientists check and recheck each other's results. Test, test, and verify. The more impactful and controversial those results, the more important is the need for verification. In this case, the process of science worked, though the process of getting the answer right took a long time.

In 1961, a young Carl Sagan jumped into the debate about water on Mars. After noting that, "To date, all spectroscopic searches for water vapor on Mars have been negative," he carried out a set of calculations to determine how much water might exist in the Martian environment, consistent with these nondetections. He concluded that the polar caps might be only 1 millimeter thick with water ice and that the amount of water vapor in the atmosphere might be nearly nonexistent. Nevertheless, Sagan suggested optimistically that "these low water-vapor abundances do not argue against life on Mars; obligate halophiles* are known which obtain their entire water requirements from the water absorbed on a crystal of salt."[12]

Finally, in April 1963, a full century after Huggins made the first spectroscopic attempt to detect water vapor in the atmosphere of Mars, two different research teams, both using modern instruments and techniques, came very close to measuring the amount of water in the atmosphere of Mars. Lewis Kaplan and his co-workers Guido Münch and Hyron Spinrad obtained a believable result. What did they do differently from previous observers? They brought a bigger hammer. Using the 100-inch diameter telescope on Mount Wilson in California and a state-of-the-art, high-spectral-resolution spectrograph with new, hypersensitized emulsions on the photographic plates, they obtained a 270-minute time exposure of Mars. Even with all these advantages over their predecessors, their result, 14 ±7 microns of precipitable water vapor in the atmosphere of Mars, was marginal (the signal level of 14 was only twice the level of the background noise of 7; most scientists require a minimum detection at three times the noise level in order for a measurement to be considered a likely detection) and not everyone was convinced by this "detection."[13] Certainly, their result was definitive in placing an upper limit on the amount of water vapor in the Martian atmosphere: less than 21 microns, meaning that if all this water condensed onto the surface, it would form a layer no thicker than 21 microns (one-fiftieth of one millimeter).

Also in 1963, a team from Princeton carried out a dramatic, high-tech experiment in an attempt to measure the water vapor content of

*An obligate halophile is a salt-loving bacterium that both thrives in and requires high salt concentrations of 15–30 percent.

the Martian atmosphere. On the evening of March 1, they launched a balloon, Stratoscope II, complete with a 36-inch telescope, to a height of 80,000 feet, putting it in Earth's stratosphere, above all but 2 microns of precipitable water vapor in Earth's atmosphere. From this height, peering at Mars, their measurements would be virtually uncontaminated by terrestrial water signals. After launch from Palestine, Texas, the balloon landed in Pulaski, Tennessee, where the data tapes were retrieved for analysis. The Princeton team also used state-of-the-art detectors developed by the Texas Instrument Corporation. These special detectors, known as bolometers, were made of a gallium-doped material* that, once cooled by liquid helium to a temperature of 1.8 degrees above absolute zero, became ultrasensitive to infrared light. Using remote controls to steer an onboard television camera, the observing team sighted Mars and collected data for about forty minutes before the balloon itself floated in between the telescope and Mars, bringing a dramatic and unexpected end to the experiment.

The Stratoscope II team made a very clear detection of *carbon dioxide* gas in the Martian atmosphere. A major discovery from this project was that the amount of carbon dioxide gas in the Martian atmosphere is so high that it overwhelms any possible signals due to water vapor. But that conclusion would only emerge after a careful and thoughtful analysis of their data. First, they held a press conference shortly after the flight ended. The scientists involved in the research oscillated between appropriately cautious and wildly optimistic in presenting their unanalyzed non-results for public consumption. Team member Harold Weaver, an astronomer from the University of California, told the press, on the basis of absolutely no actual data, that "it's pretty certain" that Stratoscope II had detected water vapor. This was an early modern example of why science should not be done by press conference.

The director of the project, Martin Schwarzschild, of Princeton University said, much more wisely and cautiously, "In two weeks

---

*Doping of a laboratory-made material involves intentionally introducing impurities (in this case, the gallium) into a semiconductor (typically silicon). The impurities change the electrical properties of the semiconductor, making it more sensitive to a particular range of wavelengths of light.

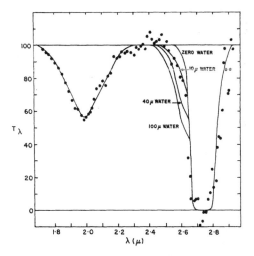

Figure 5.2. Brightness (measured as a percentage) of Mars at infrared wavelengths from 1.6 to 3.0 microns, obtained during the Stratoscope II high-altitude balloon experiment in 1963. The flat line shows the predicted spectrum of Mars assuming molecules in the Martian atmosphere absorb no light. The dots are the measured intensity of light from Mars at specific wavelengths. The lines fit through the data points are models of the Martian atmosphere. The strong, broad dips at 2.01 and 2.72 microns are due to absorption of light by Martian carbon dioxide. The dropped "shoulder" on the left side of the 2.72-micron absorption feature is due to absorption by water vapor in the Martian atmosphere. The best model fit to this shoulder indicates that Mars has about 10 microns of precipitable water vapor. The observational uncertainties in these data are apparent in the excess intensity reported from 2.4 to 2.5 microns and in the deficiency seen at 2.3 microns. Image from, R. E. Danielson, J. E. Gaustad, M. Schwarzschild, H. F. Weaver, and N. J. Woolf, *The Astronomical Journal*, v. 69, pp. 344–352, 1964, with permission.

we'll have an opinion and in three months we'll know." Nevertheless, on March 5, four days after the flight, the *Wall Street Journal* reported "Lower Life Forms May Be Able to Live in Mars Atmosphere, Balloon Findings Show." The *WSJ* went on to tell its readers that because water has been found on Mars, "some forms of lichens or moss may be present" on Mars.[14] A year later, when the Stratoscope II team published their final analysis in a paper in the *Astronomical Journal*, they concluded, very carefully and confidently, that their measurements revealed, "it is improbable that the water vapor content of Mars is greater than 40 microns." The actual best-fit they obtained in modeling their data suggested that Mars has about 10 microns of precipitable water vapor, a result remarkably consistent with that found by the Kaplan team.[15]

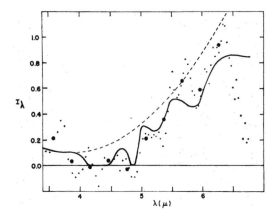

Figure 5.3. Brightness of Mars at infrared wavelengths from 3.5 to 6.5 microns, obtained during the Stratoscope II high-altitude balloon experiment in 1963. Brightness arbitrarily normalized to zero at 4.3 microns. The rising dashed line shows the predicted spectrum of Mars assuming the molecules in the Martian atmosphere absorb no light. The dots are the measured intensity of light from Mars at specific wavelengths (large data points are means, taken at 0.3-micron intervals). The solid line is a model fit to the data. The three absorption dips in the model at 4.3, 4.8, and 5.2 microns are due to Martian carbon dioxide, and the model is well matched by the data points in this spectral region. The dip in the model spectrum from 5.5 to 6.3 microns is due to Martian water and assumes 70 microns of precipitable water vapor in the Martian atmosphere. The data points in the 5.5–6.3-micron region indicate that Mars has much less water than this amount. Image from, R. E. Danielson, J. E. Gaustad, M. Schwarzschild, H. F. Weaver, and N. J. Woolf, *The Astronomical Journal*, v. 69, pp. 344–352, 1964, with permission.

While Kaplan and his colleagues along with the Stratoscope II team should perhaps share credit for the first correct (though marginal) detection of the presence of a slight amount of water in the Martian atmosphere, over the years that followed several groups carried out the first definitive and incontrovertible measurements of the amount of water. Ronald Schorn, of the California Institute of Technology and the Jet Propulsion Laboratory, summarized our knowledge of the measurements of water on Mars in a major review article written for the International Astronomical Union in 1971. "Water exists on Mars," he wrote. "The water vapor varies with location on the planet, the season on Mars, and from year to year. The water appears to cycle through the polar caps, which are partly H_2O. The total amount of water in the atmosphere of Mars is at most a few cubic kilometers."[16] If we were to spread a few cubic kilometers of water evenly across the entire surface of Mars, Mars would be covered in water to a depth of no more than about 20 microns. In other words, both Kaplan's team and the Stratoscope II team obtained the correct answer in 1963. That's not much water, but it is the right answer. Mars is very dry.

A century after Huggins and Janssen claimed to have detected water in the atmosphere of Mars, a team of astronomers who wrote a summary in 1992 of what was known about Mars 15 years after NASA's 1976-era Viking missions to Mars had wrapped up, wrote very simply and succinctly about the work of both Huggins and Janssen: "their results are no longer believed."[17]

The Martian atmosphere has a tiny bit of water vapor, but the Martian atmosphere is exceedingly dry, too dry to produce the strong spectroscopic effects seen by Huggins and Janssen. Modern astronomers are convinced that Huggins and Janssen could not have detected the tiny amount of water vapor that is actually present in the Martian atmosphere. Using the tools of astrophysical spectroscopy to prove that water is present in the Martian atmosphere turned out to be much harder than either Huggins or Janssen realized.

6

red vegetation and reasoning beings

Despite the skeptical (and correct) views of late-twentieth-century experts concerning the primitive, 1860s spectroscopic "detections" of water in the Martian atmosphere, by 1870 an overwhelming consensus had developed among the Martian cognoscenti. Mars has water and the expert astronomers of the era (seeing very clearly what they wanted to see) had proven that fact. This consensus may have been quite wrong, based as it was on the over-zealous interpretation of the data in-hand, from both images and spectroscopy. Nevertheless, astronomers in the late nineteenth century felt quite certain they had proven that Mars has considerable amounts, not just wisps, of water in its atmosphere. These astronomical results, which apparently had been confirmed by spectroscopy, meshed completely with the expectations of astronomers based on two centuries of visual observations of Mars. In fact, astronomers were certain about two important facts pertaining to Mars: it has water and it has vegetation. Having proven the former, all they had to do was prove the latter.

Based on the consensus conclusion that the Martian atmosphere was humid, Professor Huggins, or someone else with his knowledge about astronomy who was also a huge supporter of Huggins, used the pages of the relatively new but wildly successful English magazine, *Cornhill*, to turn Huggins into a celebrity (Huggins was born in

Cornhill, a traditional division of the City of London). The first edition of the monthly *Cornhill* had a publication date of January 1860. The magazine cost only a shilling and featured leading authors, including works of fiction as well as nonfiction. Published on publisher's row of Victorian London, *Cornhill* almost immediately became one of the most influential and widely read magazines of its time. The first issue sold more than 110,000 copies, and *Cornhill* remained popular and influential for decades.

In 1871, *Cornhill* published an anonymously authored piece entitled "Life in Mars,"[1] which described Mars as "a charming planet . . . well fitted to be the abode of life." With its seasons, days, clouds, and continents, "that Mars is a world like ours can be recognized most clearly by all who care to study the planet with a telescope of adequate power." After a discussion of some length about the likely oceans and seas on Mars, about the beings on Mars who might be 14 feet high because of the planet's weaker gravity, about the likely greater powers of "Martial beings, and the far greater lightness of the materials they would have to deal with in constructing roads, canals, bridges, or the like, we may very reasonably conclude that the progress of such labours must be very much more rapid, and their scale very much more important, than in the case of our own earth." Surely, though, readers are informed, proving any of this must be a hopeless enterprise, "unless the astronomer could visit Mars and sail upon the Martial seas."

But no! shouted the author. Astronomers, the most heroic of them, possess a tool, the "ally of the telescope," the spectroscope. With a spectroscope, deployed by "our most skillful spectroscopist, Dr. Huggins, justly called the Herschel of the spectroscope," Mars cannot retain its secrets. Huggins, readers learned, pulled back the curtains on Mars, once and for all. His research about Mars

> removes all reasonable doubt as to the real character as well of the dark greenish-blue markings as of the white polar caps. We see that Mars certainly possesses seas resembling our own, and as certainly that he has his arctic regions, waxing and waning, as our own do, with the progress of seasons. But in fact, Dr. Huggins's observation proves much more than this. The aqueous

vapor raised from the Martial seas can find its way to the Martial poles only along a certain course—that is, by traversing a
Martial atmosphere.[2]

From other observations, in particular those reported by Norman
Lockyer,* the author of the *Cornhill* article reported that we now
know that "the Martial mornings and evenings are misty." Also, "winter is more cloudy than summer." The article concluded by noting,
"we seem to recognize abundant reasons for regarding the ruddy
planet which is now shining so conspicuously in our skies as a fit
abode for living creatures. It would seem, indeed, unreasonable to
doubt that that globe is habitable which presents so many analo-
gies to our own." With this conclusion, the dots had been connected,
directly, explicitly, and publicly, between the discovery of water on
Mars and the assertion that life on Mars is not just probable but
likely, all because Mars has water.

The unknown author of the *Cornhill* article knew his astronomy
well. He communicated the ideas of astronomy to lay readers with
enthusiasm, and he turned Huggins into something of a folk hero,
a mid-nineteenth-century example of how astronomical discoveries
that prove the similarities of Earth and Mars lead to fame, if not fortune. He also took the cutting-edge astrophysical work of Huggins,
along with Huggins's opinions about life on Mars, and moved those
ideas from the dusty pages of a prestigious professional journal into
the living rooms of hundreds of thousands of educated Londoners.
Prior to the publicity surrounding Huggins's work, the debate about
Mars and life on Mars had been kept within the small, quiet domain
of the astronomy community. Those days were over. The life on Mars
debate, fueled by the apparent proof of the presence of abundant
water on Mars, was now part of the public conversation.

The *London Reader*, an inexpensive London newspaper, published
a short article in 1873, titled "The Planet Mars—Is It Inhabited?" in

---

*The association with the work of Lockyer was strategically wise. Lockyer had, in 1869,
founded the journal *Nature*, which quickly became and remains one of the most influential
professional science publications in the world.

which, once again, "the eminent physicist Huggins solves the problem." Mars, Huggins proved, has seas, clouds, snow, ice, fog, and rain.
"Reasoning from this basis, we can trace the presence of winds which
shift the masses of vapour from place to place, of aerial and oceanic
currents, of rivers flowing to the seas, of a climate tempered in the
same matter as our own, and of copious rainfall which must nourish
the land and cause the production of vegetation." Mars is, in fact, "a
miniature of our Earth. Here then, millions of miles away in space
is another world, a small one, it is true; but it has water, air, light,
winds, clouds, rains, seasons, rivers, brooks, valleys, mountains, all
like ours."[3] These facts, in this laundry list of assertions about conditions on Mars, all arose from Huggins's spectroscopic discovery of
proof of water on Mars. In truth, no astronomer at that time could
see winds or rain or rivers or brooks or valleys on Mars, but actual
facts were not going to get in the way of certainty about what Mars
must be like.

The author of the *London Reader* article then put forward a number of arguments related to the amount of heat Mars receives from
the Sun and the thinness of the Martian atmosphere. From these, the
author concluded that

> the weight of evidence, it seems to us, is against the existence
> of beings of a nature with which we are familiar. No terrestrial
> creature could live even in the torrid zone of Mars, so cold and
> dismal it must be. Even vegetable life, however hardy, would
> not survive a single hour. If inhabitants there be they must be
> of different form from us, to correspond to the decreased attrac-
> tion of gravity; if red vegetation exist, their eyes must be differ-
> ent from ours; to live in such an atmosphere their respiratory
> organs must be totally unlike our own.[4]

Cornhill returned to the subject of life on Mars with another anonymously authored article in 1873, titled "The Planet Mars: An Essay
by a Whewellite."[5] The title would have immediately informed readers that the author was skeptical about the idea that life could exist
on Mars. The Whewellite who wrote this *Cornhill* essay was none
other than Richard Proctor who, in his *Charts of Mars*, had named the

Martian continents and oceans. Proctor first presented arguments in support of the thesis that Mars is very different from Earth and consequently is unsuitable for life as we know it. The atmospheric pressure must be less than one-tenth that of Earth because the mass of Mars is, similarly, less than 10 percent of the mass of Earth. In addition, Mars must be much colder than Earth since it is 50 percent farther from the Sun than Earth. The clouds of Mars, the author argued, are likely cirrus rather than cumulus, and so might not be rain clouds. The author concluded that "Mars is quite unlike the Earth, and unfit to be the abode of creatures resembling those that inhabit our world. Neither animal nor vegetable forms of life known to us, he argued, could exist on Mars. Our hardiest forms of vegetable life would not live a single hour if they could be transplanted to Mars." Such conclusions sound depressing for those who imagined that life exists on Mars. Yet, the author drew exactly the opposite conclusion: "Life, animal as well as vegetable, there may indeed be on the ruddy planet. Reasoning creatures may exist there as on the Earth." Martian life will simply be different from terrestrial life, so different that "to reasoning beings on Mars, the idea of life on our Earth must appear wild and fanciful in the extreme, if not altogether untenable." If this very knowledgeable astronomer and skeptical Whewellite had capitulated, forced by the apparently overwhelming evidence to conclude that Mars is populated by both animals and plants, then the scientific debate was essentially over.

Meanwhile, on the continental side of the English Channel, via the popular writings of Camille Flammarion, the idea that Mars was not only Earthlike but likely inhabited by some kind of living things, even if perhaps different from us, became deeply rooted in the minds of many Europeans in the latter part of the nineteenth century. In 1873, Flammarion followed up his own 1869 and 1871 studies of Mars with new observations, which he presented to the Academie des Sciences in Paris in July. He described to the members of the French Academy "a polar sea around the north pole," which he was able to identify as such "because a dark patch is constantly visible there." Flammarion's polar sea extended from 80° north, where it touched the polar ice, to as far south as 45° north latitude. There, this "long, narrow Mediterranean" joined "a vast sea which extends from beyond the equator into the southern hemisphere." Flammarion pointed out an important way, in his view, in which Mars differed from Earth: on Earth, "three-quarters of the globe is covered with water; with Mars, on the other hand, there is more continental than maritime surface."[6] "Nevertheless," he continued, "evaporation on Mars produces effects analogous to those in terrestrial meteorology," and, referring now to the work of Huggins and Janssen, "spectral analysis shows that the atmosphere of Mars is charged with *water vapour*, as is ours; also that the seas, snows, and clouds are composed of the same water as our own seas."

Flammarion then felt compelled to speculate about the red color of Mars. He and others had by now developed arguments that allowed them to attribute this color to the surface rather than to the atmosphere of Mars. Flammarion took these arguments one step further: "Since it is the surface which we see, not the planets [*sic*] interior, the red colour ought to be that of the Martian vegetation, since it is this species of vegetation which is produced there." The continents of Mars, he concluded, "seem to be covered with reddish vegetation."[7]

A debate ensued over the following year among some members of the French Academy as to whether the near-invariability of the reddish color of Mars was an indication of the presence of life or proof its absence. One Academy member, Dr. Hoefer, argued that the nearly constant color of Mars was a strong argument in favor of the absence of life; soil, he argued, does not vary in color with the changing seasons. Others, like Flammarion, argued that the doctrine of the plurality of worlds offered a strong argument against the sterility of Mars. A sterile planet, especially one as capable of hosting life as Mars clearly is, would be "contrary to all known effects of the forces of Nature. There must be *something*," he argued, "on the lands, whether it be moss or even less." The arguments in favor of life on Mars were easy to muster: Mars has "species of vegetation which do not change [color with the seasons]," in the same way that on Earth "olive-trees and orange-trees are as green in winter as in summer."

By 1879, Flammarion had become a popularizing voice of astronomy on both sides of the Atlantic. In an article he authored in the *Scientific American Supplement*, titled "Another World Inhabited Like Our Own," he explained that he was motivated by his "persistent desire to find in practical astronomy a direct demonstration of this great truth of the plurality of worlds."[8] From his perspective, the detection of water in the atmosphere of Mars by Huggins and Janssen had been a game changer. "Water vapor identical with that which produces our fogs, our clouds, and our rains" had been discovered on Mars by "spectrum analysis," he explained. With this new knowledge, Flammarion asserted that astronomers now understood that the best time to study Mars, if we are to learn what is happening on the surface, would be when the atmosphere was not filled with clouds. "In other words, it must be while the inhabitants of the latter planet are enjoying fine weather." Flammarion explained to his readers that the Martian seas "are Mediterranean" with the color of Martian water "the same as that of terrestrial water." As for the red colors of the continents, "can it be," he wrote, "that this characteristic color of Mars ... is due to the color of the grass and other vegetation which must cover its plains? Can there be red meadows and red forests up there?" Putting all the pieces of information about Mars together, Flammarion asked, "Is the red indeed *terra firma*, is the green really water, and is the white indeed snow?" "Yes!" he excitedly concluded.

Nineteenth-century ideas on the evolution of planets also led to interesting ideas when comparing Earth and Mars. Astronomers insisted that Mars was older than Earth. According to a then increasingly popular theory of cosmic evolution, one based on absolutely no physical evidence but a great deal of imaginative speculation, as a planet ages the seas are progressively absorbed by the solid, planetary core. Slowly but surely, according to this (now discarded) theory of cosmic evolution, a planet loses its oceans to its own interior while simultaneously the surface dries into desert. Thus, according to then-current theory, the observational facts of Mars reveal it to be a dying planet, far closer than Earth to the end of its lifetime.

Together, a handful of astronomers transformed their beliefs about Mars into established Martian facts. They proved that Mars has water; they proved that the red patches on Mars were vegetation. They proved what they wanted to prove, and they didn't let any of their observational data get in their way. Thus, a wet, lush planet became the intellectual legacy about Mars that would inform the next generation of Mars professionals and enthusiasts.

7

water on mars: the real deal

How much water does Mars have today? A century and a half after William Huggins first "proved" that Mars had water, astronomers have now solved some, but certainly not all, of the riddles of how much water Mars had and still has. The simple answer is that Mars has lost most of the water it once had. Nonetheless, Mars still has a great deal of water, just not in the liquid state on the surface anymore. Martian surface features show, very clearly, that Mars had enormous quantities of water flowing on its surface during the first half-billion years of Martian history, until as recently as 3.9 billion years ago. Thereafter, the surface of Mars became much dryer, either because the water disappeared below the surface, where it might remain, or escaped into space, or both.

Significant evidence has emerged over the last two decades, however, that strongly suggests Mars also experienced a more recent epoch when water-carved valleys, probably formed from melting snow and ice, flowed slowly into and out of chains of lakes. Some of these fluvial features, identified in images obtained by NASA's Mars Reconnaissance Orbiter and Mars Global Surveyor and ESA's Mars Express spacecraft, may be as old as three billion years; others may have formed as recently as two billion years ago. A single one of these lakes, which appear to have been widespread both north and south of the Martian equator, contained more water than Lake Ontario.[1]

Mars has layered ice deposits at both the north and south polar caps. Some of this ice is permanent (in the current climate conditions of Mars), but a thin layer is deposited each Martian winter and then sublimates in Martian spring. In 2001, David E. Smith, of the Goddard Space Flight Center, and Maria T. Zuber and Gregory A. Neumann, both from the Massachusetts Institute of Technology, used laser altimeter data obtained by Mars Global Surveyor, which entered Mars's orbit in September 1997, to show that the seasonal 1.5–2-meter-thick (5–6.5 feet) veneer of ice that covers the north polar cap in winter is frozen carbon dioxide (dry ice) and that the polar cap that remains through all of northern summer is water ice.

Primarily because it lies at an elevation 4 miles higher than the northern polar cap, the southern polar cap retains a thin veneer of dry ice over its permanent water ice cap, even in southern summer.[2] From this work, we know that almost all of the volume of the ice caps, below the veneer, is water ice and that most of the water that remains *on the surface* of Mars is probably stored in these polar caps.

In 1998, a science team led by Zuber and Smith had used the Mars Global Surveyor laser altimeter to estimate the total volume of ice contained in the *northern* polar cap. Four years later, their work revealed that they could confidently reinterpret the 1998 results as measurements of the total volume of *water* ice in the northern polar cap. The answer: those measurements yielded a cap volume of between 300,000 and 400,000 cubic miles of water ice (about half the volume of the Greenland ice cap). This is enough ice such that, if melted, it would produce a global ocean, that being an ocean covering 100 percent of the Martian surface, about 30 feet deep.[3]

The Mars Express spacecraft, launched in 2003, entered Martian orbit in December 2003. Jeffrey J. Plaut, of the Jet Propulsion Laboratory, used the Mars advanced radar system on Mars Express to measure the total ice volume contained in the *southern* polar cap. The answer: 400,000 cubic miles of water ice, which is enough water to create a global ocean with a depth of about 36 feet.[4] Together, the northern and southern polar ice deposits contain enough water to generate a global ocean with a depth of about 66 feet.

In 2013, Jeremie Lasue, of the Université de Toulouse, in France, and his colleagues summarized all of the then-known near-surface water reservoirs on modern-day Mars, as measured by numerous satellites and instruments over the last two decades.[5] Some water is known to be stored in the northern and southern polar deposits, while additional water is found in permafrost layers outside of the polar caps, in near-surface deposits scattered around the globe. In total, they found that Mars was known to have enough water in these near-surface reservoirs to form a global ocean between 80 and 96 feet in depth.

In 2015, Geronimo Villanueva and Mike Mumma, of NASA's Goddard Space Flight Center, made their own determination of the amount of water Mars has now and has *lost* to space over time. In order to understand their work, we need to remember a bit of chemistry.

Water is often identified by its atomic formula, H_2O, which is a chemist's shorthand for identifying the elemental constituents of a single water molecule as two atoms of hydrogen (H) and one atom of oxygen (O). Not all hydrogen atoms are identical, however. Deuterium is an *isotope* of hydrogen. Like normal hydrogen, a deuterium atom has one positively charged proton in the nucleus and a positive nuclear charge of +1; unlike normal hydrogen, however, deuterium also has a neutrally charged neutron in the nucleus. Neutrons, whose masses are nearly identical to the masses of protons, increase the masses but do not affect the charges of atoms; thus, adding one or more neutrons to an atomic nucleus does not impact the chemical behavior of the atom. All of this means that a deuterium atom is simply a heavy form of hydrogen.

If one of the hydrogen atoms in a water molecule is replaced with a deuterium atom, that molecule (HDO) is still water, but we refer to it as "semi-heavy water." If we replace both hydrogen atoms with deuterium atoms, the molecule (D_2O) is known as "heavy water." Thus, a water molecule can be H_2O or HDO or D_2O. All three would taste the same, but they would have different masses and different weights, respond differently both physiologically and under the influence of gravity, and absorb and emit light at slightly different wavelengths.

On Earth, one out of every 6,400 hydrogen atoms exists in the form of deuterium. As a result, about one out of every 3,200 water molecules is actually a semi-heavy water molecule rather than H_2O.

Also, about one in every 41 million water molecules ($6,400^2 = 41$ million) on Earth is naturally a heavy water molecule.

Using telescopes in Hawai'i and Chile, Villanueva and Mumma measured water in the atmosphere of Mars in two different forms: normal water and semi-heavy water. In the Martian environment, ultraviolet light from the Sun will break water molecules into their atomic constituents, hydrogen and oxygen. Since hydrogen is very light in comparison to oxygen, it will bubble up toward the top of the atmosphere and escape into space, while the oxygen settles to the surface and reacts with minerals in rocks to form rust. Normal hydrogen, having only half the mass and weight of deuterium, eludes the gravitational grasp of Mars more easily and escapes to space significantly more quickly than deuterium. As a result, very slowly but steadily over the eons, the ratio of deuterium to hydrogen (D/H) increases* as water molecules in the atmosphere are broken apart by ultraviolet light and some of the hydrogen and deuterium atoms escape to space. The D/H ratio on Mars today is now eight times greater than the D/H ratio on Earth.

Because both Earth and Mars formed out of the same cloud of gas that surrounded the Sun, planetary scientists believe that both planets should have started with identical D/H ratios (unless the water on both planets was delivered by comets** long after the surfaces of the planets had cooled). In addition, the magnetic field and larger mass of Earth act together to prevent the destruction and loss of water from Earth. Consequently, the D/H ratio on Earth today should be identical to the D/H ratio on both Earth and Mars when the two planets formed, more than four billion years ago.

Assuming that the current surface reservoir of water on Mars is equivalent to a global layer of depth 69 feet (perhaps an underestimate of 10–30 percent, based on Lasue's estimates), and assuming the current D/H ratio on Mars is the result of the photodissociation of some Martian water and the subsequent loss to space of normal hydrogen atoms more quickly than deuterium atoms, Villanueva

---

*The amount of H decreases faster than does the value of D, so the ratio D/H increases over time.

**Comets are enriched in deuterium relative to hydrogen and have a wide variety of D/H ratios.

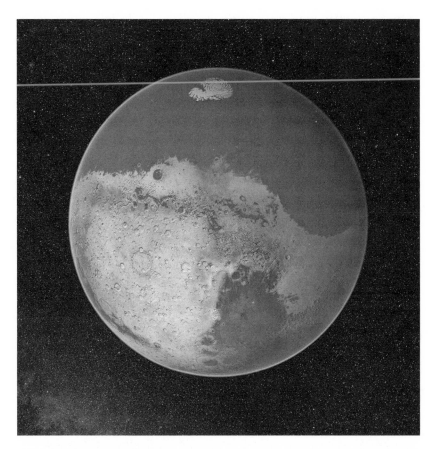

Figure 7.1. Conceptual map showing a large ocean, holding more water than Earth's Arctic Ocean, on ancient Mars, as reported by Villanueva et al. in *Science* in 2015. Given current Martian topography, which likely has not changed significantly in billions of years, the ocean would be in the low-elevation Northern Plains region. The idea that Mars once had an ocean emerges from measurements that indicate Mars has lost 85 percent of the water it once had to space. Note that if ocean basins once existed on Mars, they would have filled enormous, ancient impact craters, whereas ocean basins on Earth are the result of plate tectonics, which is and has always been absent on Mars. See Plate 3. Image courtesy of NASA's Goddard Space Flight Center.

and Mumma calculate that Mars once had 6.5 times more water than it has today, or enough water to create a global ocean with a depth of 450 feet. Thus, according to Villanueva and Mumma, the current D/H ratio on Mars proves that Mars has lost about 85 percent of the water that has been on or above the surface and therefore vulnerable to destruction by ultraviolet light.[6]

Based on what we know now about D/H ratios on Earth and Mars, about the formation of Earth and Mars, and about the amount of water near the surface of Mars, Villanueva's and Mumma's conclusions are reasonable. We should, however, recognize that our knowledge about all of these things remains incomplete, and so conclusions about how much water Mars has now and has lost over time also can only be considered plausible, not definitive. In fact, by many estimates, much of the water needed to account for some of the surface features of Mars remains unaccounted for in the current inventory of water ice combined with Villanueva's and Mumma's projection of water abundance on ancient Mars based on the evidence of atmospheric loss of 85 percent of the surface reservoir of water.

Some noteworthy evidence that Mars once had massive amounts of water is the presence of ancient valley networks, which are similar to river valley networks on Earth. The largest of these water-carved valleys are as much as half a mile in width and a thousand feet in depth. Clearly, an enormous amount of water flowed through these networks when they were active, more than 3.7 billion years ago. The amount of water needed to carve out these valley networks is probably equivalent to a global ocean with a depth of a thousand feet.

Mars also has surface features known as outflow channels. These channels are fossils of moments in Martian history when enormous volumes of water were released catastrophically, perhaps when permafrost layers near the surface melted quickly. All that meltwater subsequently spilled onto the surface, where it washed downhill and carved out these deep channels. Some of these outflow channels are tens of miles wide and 600 miles in length. The discharge rates necessary to have formed some of these outflow channels are as much as one hundred times greater than the flow rate of the Amazon River. On Mars, the identified outflow channels require enough water to generate a global ocean that would have been 1,500–3,000 feet deep.[7]

While such massive outflows seem hard to comprehend, one similar natural catastrophe occurred in recent history on Earth. About 12,000 years ago an enormous glacial lake, Lake Missoula, containing as much water as is found today in Lake Erie and Lake Ontario

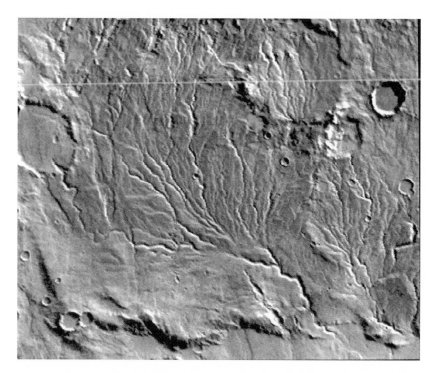

Figure 7.2. Valley network (location 42°S, 92°W), which resembles a natural drainage system on Earth, reveals evidence of water flowing over long periods of time on the surface of ancient Mars. Area shown is about 120 miles across. Channels merge together to form larger channels. The lack of small-scale streams feeding the larger valleys suggests these valleys were either carved by groundwater flow, perhaps under a layer of ice, rather than by runoff from rain, or that Mars was once warmer and wetter than it is today. Image courtesy of the Lunar and Planetary Institute/Brian Fessler.

combined, filled a series of valleys to the east of a 2,500-foot tall ice dam in Idaho. Eventually, the enormous water pressure caused a catastrophic failure of the dam. The volume of the Lake Missoula outflow has been estimated to be about sixty times that of the Amazon, with water cascading at rates of up to 50 miles per hour, carving out valleys and leaving behind evidence of this catastrophe in the form of buttes and deep canyons across eastern Washington. At this flow rate, Lake Missoula likely drained in about a week.[8]

The amounts of water that must have been present on Mars in the distant past are far more than the global equivalent layer of 450-foot depth estimated from the water and gas inventory in the atmosphere

Figure 7.3. A 200-mile-long portion of the Ravi Vallis outflow channel (location 1°S, 42°W). The channel begins at the left side of the image. The absence of any tributaries suggests that the water that carved this channel was released from an underground reservoir at great pressure. The rapid flow disrupted, collapsed, and scoured the surface. Image courtesy of the Lunar and Planetary Institute/Brian Fessler.

and at the surface now. Either most of Mars's water was lost to space through a mechanism that did not affect the D/H ratio, or most of Mars's water remains on the planet, stored well beneath the surface. The latter possibility, if we can find and access the hidden reserves of water, makes the possibility of colonizing and terraforming Mars plausible.

In 2016, scientists making ground-penetrating radar soundings with the Mars Reconnaissance Orbiter spacecraft discovered an ice sheet in western Utopia Planitia (a large plain in the northern hemisphere of Mars, located inside the largest impact basin both on Mars and in the solar system) that is several hundred feet thick, covers an area larger than New Mexico, and holds more water than is found in Lake Superior.[9] Undoubtedly, we have not yet found all the water on Mars. These measurements of the water ice content of the polar caps and in at least one nonpolar ice sheet provide a preliminary answer to the question, "How much water does Mars have today at or near the surface?" We need additional data to discover whether Mars has retained all of the water it once had.

In 2015, NASA's Mars Atmospheric and Volatile Evolution mission, MAVEN, provided an important answer concerning the history of water on Mars. MAVEN was launched in 2013, arrived at Mars in 2014, and since then has been working to measure the rate at which the Martian atmosphere loses atoms to space. Some atoms are knocked off the highest, thinnest layer of the atmosphere and out into space when a particle from the solar wind collides with one of those atoms. The solar wind consists of the charged nuclei of hydrogen and helium atoms that stream away from the Sun at speeds of about 250 miles per second. The solar wind particles not only travel outward from the Sun, they grab and stretch the Sun's magnetic field lines out into the solar system. In doing so, they extend the influence of the Sun's magnetic and electric fields out to the planets. The magnetic field of Earth pushes back against the onrushing solar wind particles and forms a shield, well above the top layer of Earth's atmosphere that protects the upper atmosphere from being eroded by the solar wind. Mars, however, has no magnetic field. As a result, the upper layer of the atmosphere of Mars is exposed to the sandblasting effect of the solar wind. In addition, some atoms in the upper Martian atmosphere absorb sunlight. As a result, they are energized enough to lose one of their electrons. When missing an electron, that atom is now a charged particle, an ion. Ions react and respond to magnetic fields, whereas neutral particles do not notice the presence of a magnetic field. When ions from the atmosphere of Mars come into contact with the Sun's magnetic field, they are yanked outward and away from Mars.

Slowly and steadily, second by second, day after day, year after year, these two processes (first ionization, then interaction with the solar wind) together cause Mars to lose parts of its atmosphere at the rate of about a few hundred grams (about half of one pound) per second. "Like the theft of a few coins from a cash register every day, the loss becomes significant over time," said Bruce Jakosky, MAVEN principal investigator at the University of Colorado, Boulder. "We've seen that the atmospheric erosion increases significantly [note: as much as ten times greater] during solar storms, so we think the loss rate was much higher billions of years ago when the Sun was young and more active."[10] Atmospheric loss is also as much as ten times greater when Mars, which follows an elliptical orbit, is closest to the Sun than when Mars is farthest from the Sun.[11]

The MAVEN science team has also used measurements of the relative abundances of two isotopes of argon at different heights in the atmosphere to determine the amount of gas lost over time from the Martian atmosphere to space.[12] As with deuterium and normal hydrogen, the gravity of Mars is very slightly better at holding on to the heavier isotope of argon (Argon-38, or ^{38}Ar) than the lighter isotope of argon (Argon-36, or ^{36}Ar). The measurements reveal that Mars has lost 66 percent of its initial reservoir of argon to space. The MAVEN scientists were able to use the lost argon measurements to estimate that "as much as half a bar or more of CO_2 could have been removed" from the atmosphere of Mars through time. One "bar" is the equivalent of the surface pressure of the atmosphere on Earth (about 14.5 pounds per square inch). As the surface pressure on Mars today is one hundred and sixty times smaller than one bar, these MAVEN measurements provide strong evidence that a very large fraction of the gases (e.g., argon, molecular nitrogen, carbon dioxide) that once resided in the atmosphere of Mars have escaped to space.

Given that the MAVEN measurements make a very strong case that Mars has lost (and continues to lose) most of its original atmosphere, we want to ask how that helps us better understand how much of the original Martian reservoir of *water* has been lost over the lifetime of Mars. One tool for making such an estimate is the neutron spectrometer instrument on NASA's Mars Odyssey spacecraft. This detector measures the flux of neutrons emerging from within three feet of the Martian surface. These neutrons, in turn, are generated by the collisions of high-energy particles from space, known as cosmic rays, which penetrate the Martian surface and hit near-surface particles. When cosmic rays collide with atomic nuclei, the collisions generate gamma rays and neutrons, some of which ricochet back up out of the atmosphere, where they can be detected by Mars Odyssey. Scientists can use the energies of the neutrons to identify what particles exist just beneath the surface of Mars.

One of the big discoveries by Mars Odyssey is that the Martian subsurface at high northern and high southern latitudes (both 55 degrees north and south, and closer to both poles) is extremely rich in hydrogen. Since hydrogen would not exist by itself in rocks near the Martian surface, the most likely reason for the presence of hydrogen is that water ice is abundant in the subsurface rocks. That is, the Martian poles are dirty ice, known as permafrost, which is likely to be about 50 percent rock mixed with as much as 50 percent water ice by volume.[13] Equatorial regions also contain water, but at the lower level of 2–10 percent by volume.[14] NASA's Phoenix Lander mission landed near the north pole of Mars in 2008 and used a robotic arm to scoop up Martian dirt and measure its contents. Phoenix showed that Odyssey was right: the dirt at high northern latitudes is full of water ice.

According to the MAVEN and Odyssey and Phoenix measurements, over the last four billion years, Mars may have lost the equivalent of an ocean of water. For Mars to have had this much liquid water on or near its surface, it would have had to have been much warmer than it is today. The Mars Reconnaissance Orbiter has uncovered evidence, using radar to probe about half a mile beneath the surface of the south polar cap, of enormous, layered reservoirs of frozen carbon dioxide. Thin layers of water ice (each 50–200 feet thick) separate three thick layers of frozen carbon dioxide (each about 1,000 feet thick). The total amount of carbon dioxide locked into these deposits is nearly equal to the total amount of carbon dioxide currently in the Martian atmosphere. If released, the pressure in the Martian atmosphere would double, which would permit liquid water to be stable in many locations on the surface of Mars. In addition, with a denser atmosphere, stronger winds could loft more dust into the atmosphere, which likely would lead to an increase in the frequency, and perhaps the intensity, of Martian dust storms.[15]

On that ancient warm and wet Mars, sedimentary layers made from clays and minerals such as hematite would have formed, and indeed they have now been identified. First spotted in 1998 near the Martian equator by Mars Global Surveyor,[16] these hematite layers provide evidence for large-scale water interactions on or near the surface of a young Mars.

Where is all of the water Mars once had? Mars definitely lost some and possibly a great deal of its original inventory of water. But if it did so, much of that water escaped without affecting the D/H ratio. Quite possibly, the water is still on or inside Mars.

8
---

canal builders

canali

In the beginning, Giovanni Virginio Schiaparelli was not even interested in Mars, let alone in looking for Martians. Yet by the time he had finished his studies of Mars, he had redrawn and re-imagined the surface of Mars and discovered what he believed to be a planet-girdling system of 75-mile-wide canals.

Prior to making his revolutionary observations of Mars, Schiaparelli already had earned worldwide fame as an extraordinarily good astronomer. In 1866, four years after two American astronomers, Lewis Swift and Horace Tuttle, discovered the comet that would one day bear their names, Schiaparelli calculated that the orbit of Comet Swift-Tuttle was similar to that of the meteoroids that, every August, produce the meteor showers known as the Perseids (because they seem to come from the direction of the constellation Perseus). In connecting the "shooting stars" produced in these meteor showers to the orbit of a comet, Schiaparelli connected the physics of the dusty debris swept up by Earth as it orbits the Sun to the detritus shed by comets as they orbit the Sun. For this major astrophysical discovery, in 1868 the French Académie des Sciences bestowed on Schiaparelli the premier international prize in astronomy at the time, the Lalande Prize.

Schiaparelli had already benefited from being in the right place at the right time. When in his early twenties, the patronage of the king of Sardinia helped secure for him government support to train first in Germany at the Berlin Observatory and then in Russia at the Pulkova Observatory (located outside of Saint Petersburg). Afterward, he was given a patronage appointment as the assistant astronomer at a small observatory in Milan. At age twenty-seven, in 1862, he was promoted to director of Brera Observatory when the senior astronomer died. Schiaparelli immediately put forward an ambitious modernization plan, including a proposal for a state-of-the-art telescope. He made little progress for the first few years, but receiving the prestigious Lalande Prize helped Schiaparelli solidify the political patronage that paid his bills. Also by 1868, the King of Sardinia, whose ministers had helped young Schiaparelli get started in his career, had been crowned Vittorio Emanuele II, king of newly (since 1861) united Italy. Schiaparelli's scientific success had potential as a good public relations investment for King Emanuele II's young country, and so the king chose to fund Schiaparelli's request for a new telescope for Brera Observatory. Six years later, Schiaparelli's new telescope, the 8.6-inch Merz refractor, made with some of the highest quality optical glass that could be manufactured at the time, was ready for him to use in his astronomical observations, most of which were planned as studies of the separations and orbits of double stars.

Merely as a means of testing his new telescope, Schiaparelli conducted his first observations of Mars in the summer of 1877. Mars was positioned well in the nighttime sky and was approaching opposition. It would be at its closest approach to Earth in 12 years in September 1877, so it was already quite large when viewed through his telescope. Mars almost instantly captivated Schiaparelli. He found that he could easily pick out different markings on the surface, and he was able to discern the many shades and colors of these surface features, the history of which he was well aware from the work of other astronomers who had mapped the surface of Mars.

By September, Schiaparelli resolved to take advantage of the proximity of Mars and the quality of his new telescope to map its surface in far greater detail than anyone had done before. He even invented a technique that guaranteed he would have a more precise map of the surface than those obtained by all previous observers. His plan was to identify sixty-two fixed points on the Martian surface and then use a micrometer to measure the angular distances from these points to every other feature he wished to accurately locate on the surface of the planet. Since the best previous chart of Mars, drawn in 1867 by Richard Proctor, lacked names for most of the locations that were merely the starting points for Schiaparelli's work, let alone names for all the additional features he would soon identify, Schiaparelli would soon introduce his own nomenclature for almost every surface feature on Mars (some of which are still in use).

The result of Schiaparelli's efforts was the most accurate map yet made of the surface of Mars, at least insofar as the features observed by Schiaparelli were real. He not only described the dark areas of Mars, without reservation, as seas, he explained the differences in their color tones as due to differences in "depth, transparency, and chemical composition," just as "the difference in salinity between terrestrial seas causes the differences between the colours of these seas. The saltier the water, the darker it appears.... It is the same on Mars.... All this leads us to regard the Martian seas as similar to the terrestrial ones."[1]

He then discussed "the complicated réseau of dark lines which links the patches we regard as seas." These lines, he explains, "owe their colour to the same cause as with the seas, and can only be canals or communicating straits." These dark lines continued to fascinate Schiaparelli during his observing campaigns in 1879 and again in the fall of 1881 and early winter months of 1882. Then, in 1882, Mars apparently suddenly and dramatically changed almost in real time as Schiaparelli watched.

"On the planet," Schiaparelli wrote, "crossing the continents there are a large number of dark lines to which I have given the name *canals*, though we do not know that this is what they are. Many observers have previously recorded some of them, most notably Dawes in 1864. During these last three oppositions, I have made a special study of them, and I have recognized a considerable number—at least sixty." At least sixty! These canals form "a well-defined réseau on the bright or continental regions," he wrote. "Their disposition appears invariable and permanent, at least insofar as I can judge after having observed for 4½ years."

At this point, Schiaparelli began to break new barriers in Martian science. He reported that he found some canals in 1879 that he had not found in 1877 and found more in 1882 that he had not found in 1879, but he always recovered those found in previous years. In his professional opinion, the Martian canal system was steadily growing more extensive across the planet. The shortest was 75 miles in length, the longest 3,000 miles, and every canal "terminated, to either end, in a sea or in another canal; there is not a single example of a canal stopping short in the middle of a land-mass." By his estimates, the widths of the canals were enormous, typically 75 miles (two Martian degrees).

Then Schiaparelli brought down the hammer: "This is not all. At certain seasons these canals are doubled, or, more accurately, double themselves." In 1877, Schiaparelli had not yet detected this phenom-enon, in which a canal almost overnight became two parallel canals. He also claimed that it had only occurred once in 1879, when he "noted the doubling of the Nile." In 1882, however, Schiaparelli "had surprise after surprise; in succession the Orontes, Euphrates, Phison, Ganges, and the majority of the other canals, showed up as clear and uncontestable doubles." Schiaparelli believed he had witnessed at least twenty canals double, and seventeen of them had done so in a single 30-day period.

"These doublings are not an optical effect depending on the in-crease of visual power, as happens in the observation of double stars, and it is not due to the canal being divided into two along its length. This is what is presented: To the right or left of an existing line, with-out any change in the course or position of the line, we see the ap-pearance of another line equal and parallel to the first, at a distance ranging from 6° to 12°, that is to say 350 to 700 kilometers [220 to 440 miles]." Schiaparelli made astronomical reference to the constel-lation Gemini when he imaginatively decided to call this process, in which the canals doubled along their lengths, gemination, meaning "twinning" in Latin.[2]

Schiaparelli's new results were received, to put it mildly, with a great deal of skepticism by other astronomers. Some argued that

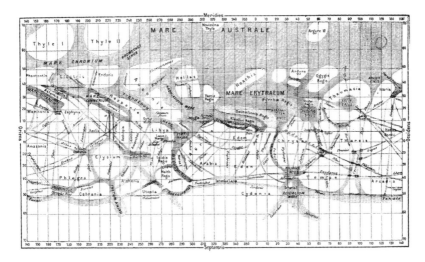

Figure 8.1. Giovanni Schiaparelli's 1886 map of Mars, showing canal structure on Mars, including many double canals, in which all canals appear to terminate at a major landmass. Image from Flammarion, *La Planète Mars*, 1892.

Schiaparelli's sharp lines were the borders of lightly shaded surface regions rather than canals. Others suggested the canals were optical illusions, the result either of bad optics in Schiaparelli's telescope or in his own head. Schiaparelli remained confident, however, that he was an astronomer both ahead of his time and ahead of his peers in making discoveries about Mars.

A few astronomers fell in line on Schiaparelli's side of the scientific battleground. They offered him support by claiming confirmation of the existence of at least some of Schiaparelli's canali. Irishman C. E. Burton provided the first confirmation of Schiaparelli's canals when he reported, in 1882 in the *Scientific Transactions of the Royal Dublin Society*, that he had identified several canals. Burton, however, was unable to confirm the phenomenon of gemination.[3] The obituary published in the *Astronomical Register* for Mr. Burton, who died suddenly of heart disease in 1883, at age thirty-five, reported that, "our knowledge of the recently-discovered Martial canals is mainly derived from observations of M. Schiaparelli and Mr. Burton."[4]

Within a few years, important additional support for the reality of Schiaparelli's canali appeared with the work of the French team

of Henri Perrotin and Louis Thollon. In 1886, working at Nice Observatory, Perrotin and Thollon even achieved the breakthrough
that Schiaparelli so desperately hoped for: confirmation of the doubling of the canals on Mars. "By the end of that night [April 5, 1886],
under good conditions, we had been able to recognize successively,
several canals presenting, in nearly all respects, almost the character
attributed to them by the Director of the Milan Observatory. These
canals are described by Schiaparelli, and as we have seen them, make
up in the equatorial region of the planet a réseau of lines which seem
to follow the arcs of great circles."[5] Perrotin and Thollon even mention that a visitor to the observatory, M. Trépied, was the first "to
have noted the two shaded parallel lines which make up the double
canal TU." Perrotin was able to see both single canals and double
canals during his next observing campaign in 1888.

Even more support that appeared to verify the work of Schiaparelli came from another French astronomer, François Terby. "We
find," he wrote in his *Physical Observations of Mars*, "that in 1888 we
have verified at Louvain the existence of the following canals." He
then lists the names of thirty canals. Terby explains how he observed
some of the canals multiple times and that some of the observations
"though difficult, did not leave the slightest doubt." Terby was disap-
pointed to have been able to confirm the doubling of only one canal;
nevertheless, he did "get a glimpse of the doubling of one of them,
in spite of the combination of deplorable circumstances." Terby concluded that "After what we have seen we dare to affirm that henceforth the progress of areography will be in the hands of those alone
who, freeing themselves from the shackles of doubt, will resolutely
engage in the way traced by the celebrated astronomer of Milan: A
new era is begun in the study of Mars by the study of canals and their
doubling."[6]

Perhaps the most important verification of Schiaparelli's work
was that offered by William Henry Pickering, the younger brother
of Edward Charles Pickering. Edward was one of the most powerful
and influential astronomers in the world. He had been appointed
as director of the Harvard College Observatory in 1876 and would
remain in that position until 1920. Young William got off to a successful start in his career as an astronomer. In the 1880s, he was

making important contributions to astronomy and was quite highly respected. He was elected as a Fellow of the American Academy of Arts and Sciences in 1883 and discovered Phoebe, one of Saturn's moons, in 1899. Later in his career, however, William would carve out a rather eccentric path, detecting (or so he thought) vegetation and giant insects on the Moon and devoting much of his energy toward searching for the mysterious planet beyond Neptune, Planet X.

In 1887, as a bequest from the Uriah Boyden family, Edward Pickering had acquired funds for Harvard Observatory to establish a permanent, high-altitude observing facility. Using the Boyden funds, Edward had an observatory built in the Andes mountains of Peru, at Arequipa, at an elevation of 8,100 feet. He assigned to William the task of setting up the Boyden observing station. William suggested to Edward that Harvard Observatory should use the Arequipa site to conduct a large-scale photographic survey of the southern sky. Though William was not involved in the later analysis of the thousands of photographic plates that were obtained at Arequipa and then shipped back to Boston, the results of this survey completely revolutionized astronomy in the twentieth century, leading to the spectral classifications of stars by Annie Jump Cannon, the discovery of the Period-Luminosity relationship for Cepheid variable stars by Henrietta Leavitt, and the creation of the Hertzsprung-Russell diagram, all of which together provide the foundation for all of modern astrophysics. William Pickering's imagination and foresight in this endeavor merit enormously high praise. His subsequent studies of Mars were much less praiseworthy.

In 1890, at Arequipa, William used the 12-inch Boyden telescope to take some of the first-ever photographs of Mars, again putting him at the cutting edge of late-nineteenth-century astronomy. In doing so, he "first saw the so-called 'canali' of Mars, discovered only a few years before by the great Italian observer Schiaparelli."[7] In 1890, he reported in an article in *Sidereal Messenger* that he had seen a great many of these canals. Some were easy to see. Others were "readily seen." More were found "even when the seeing is only moderate," as he had "no difficulty in also seeing Styx, Fretum Anian, and Hyblaeus. Several other canals in this same region have been recognized." He had "not yet been able to double any of them," and he now had "the highest admiration for the eyesight of the astronomer who could discover them in the first place with an 8-inch telescope." Surely, he suggested, almost anyone could see them; however, he also offered some degree of skepticism as to whether these Martian features were properly analogs of terrestrial canals, as "it seems to me most unfortunate that the name of canals has been attached to these finer markings upon the planet, for there has not been the slightest evidence brought forward in support of the supposition that they are filled with water."[8]

After the 1892 opposition of Mars "had passed into history," William Pickering offered a summary of his observations made at Arequipa in an article in *Astronomy and Astro-Physics* published in December of that year.[9] With specific regard to the canals, he made clear that "Numerous so-called canals exist upon the planet, substantially as drawn by Professor Schiaparelli. Some of them are only a few miles in breadth. No striking instances of duplication have been seen at this opposition." Pickering also observed "clouds on several occasions" and "minute black points." "For convenience," he wrote, "we have termed them lakes." He also claimed to be able to detect snow melting near the poles. His certainty about detecting snow near the poles, his ability to detect clouds, and his preference for labeling the small black spots "lakes" seemed to contradict his wisdom in wishing that Schiaparelli had chosen terminology other than canals for the long, linear features on Mars that showed no apparent evidence for containing water.

Schiaparelli himself never allowed himself to extend his personal speculations about the canals he had discovered on Mars as far as some others would. "It is not necessary," he wrote, "to suppose them the work of intelligent beings."[10] Others were less cautious. After all, canals on Earth are artificial constructs. Canals like the ones on Mars that were large enough to be seen through telescopes from a distance of 50 million miles were without a doubt enormous, almost unfathomably mammoth, engineering feats designed by extremely capable and intelligent engineers.

On Earth, engineers completed construction of the Canal du Midi in southern France in 1681, which was the first segment of a human-constructed combination of canals and rivers that soon would connect the Atlantic to the Mediterranean; in 1825 in up-state New York, workers finished digging the Eighth Wonder of the World, the (originally) 4-feet-deep, 40-feet-wide Erie Canal, which connected the Hudson River to Lake Erie. Three decades later, in 1858, a team led by French engineer Ferdinand de Lesseps began construction of the Suez Canal and completed that incredible project in 1868. With a length of less than 125 miles and a breadth of less than a half mile, the Suez, perhaps the greatest engineering accomplishment in the history of humankind at the time it was completed, was child's play compared to the dozens of interconnected, thousand-miles-long, tens-of-miles-wide canals on Mars. Yet, over a time period of a few centuries here on Earth, human engineers were busy building a planet girdling system of canals. In fact, at the end of the nineteenth century, during the discovery phase of the Martian canals, work was under way to build the biggest canal yet built on Earth, the Panama Canal. Beginning in 1855, the Panama Railway began to haul passengers and goods across the Isthmus of Panama. Immediately thereafter, serious surveys and discussions developed concerning a possible canal to follow. A French team began excavation of the canal in 1881; the United States took over the project in 1904 and completed the canal in 1914. In the late nineteenth century, canals were on the minds of the greatest scientists and engineers on Earth; some of those scientists were not surprised to discover that intelligent aliens were already building similar structures on a neighboring planet. Certainly, those among us who were in the business of remaking planet Earth could appreciate the advanced skills of the Martians.

Astronomers, of course, were not working in a vacuum. Working at the cutting edge of science, astronomers were well aware of the canal-building efforts of the technologically advanced French and American engineers. They had canals on their minds as they observed Mars. The Italian astronomer Secchi first applied the *canale* label to Mars in the same year when, just across the Mediterranean Sea, construction began on the Suez Canal. And though Schiaparelli found his first canals in 1877, he noticed the first occurrence of the more advanced doubled form of Martian canals in 1882, the year after the French began digging in Panama.

Canals, clearly, are what intelligent human civilizations build. To Camille Flammarion, a globe-encircling network of canals on Mars provided evidence that the planet Mars was host to an extremely intelligent civilization far more advanced than our own. As a popularizer of astronomy, Flammarion had no peer before his own time and none after until Carl Sagan wrote *Cosmos*, a century later. In August 1892, Flammarion published his *La Planète Mars*, containing a summary of almost all the important observations made of Mars since the telescope had been invented, from the first observations of Fontana in 1636 through the observations of canals by many astronomers in 1890. In addition to providing, in one place, details of every important prior observation ever made of Mars, Flammarion also offered his own, imaginative interpretation of humanity's understanding of the red planet:

> It is possible that Mars is actually inhabited by a human species analogous to our own. More light-weight, no doubt; and older, and much more advanced. However, major differences exist between the two worlds. We do not yet have enough information to speculate about the possible forms of the human, animal, vegetable and other types of life on Mars. But the inhabitation of Mars by a race superior to ours seems to be very probable.[11]

intelligent martians

Enter Percival Lowell. Lowell was a wealthy Bostonian, the son of one of the most prominent families in New England. At Harvard, from which he graduated in 1876, Lowell studied under one of the most important American astronomers of the nineteenth century, Benjamin Pierce. After graduation, he left behind his astronomical interests for 15 years while he traveled to Europe, ran the family business, and then traveled more extensively, in the later years to Korea and Japan. By the early 1890s, though, he was returning to his passion. He had wrapped his imagination around Schiaparelli's canals and became intent on reinventing himself as an astronomer. By the end of the decade, he had built his own observatory and convinced lay readers around the world that Martian engineers had built a canal system in order to save their dying world.

In 1892, through his friendship with Harvard Observatory director Edward Pickering, Lowell obtained copies of Schiaparelli's maps of Mars and began studying them with great care. Then, having learned that Schiaparelli himself had given up his studies of Mars because his eyesight had become too poor for him to continue his observational work, Lowell began imagining ways in which he might get into this business of observing Mars, picking up where Schiaparelli had left off. Then, as a Christmas gift in 1893, he received a copy of Flammarion's *La Planète Mars*. Almost overnight, Lowell decided he needed to devote himself immediately and fulltime to the study of Mars. To that end, he chose to invest both his intellectual gifts and his personal fortune in building an observatory dedicated to this purpose.

After evaluating several possible sites for his telescope in order to identify a location with the best atmospheric qualities, including locations in Mexico and Algeria, Lowell installed his state-of-the-art, Alvan Clark and Sons 24-inch refractor telescope on Mars Hill in Flagstaff, Arizona, where he believed the dryness and stillness of the atmosphere would enable him to see details on the surface of Mars that others, observing from locations with lesser atmospheric qualities, would be unable to see. Lowell pioneered this idea in astronomy, which has since led to the establishment of observatories on mountaintops in Hawai'i, the Canary Islands and in Chile, where the atmospheric "seeing" is better than at other locations on Earth.

Lowell also told others that excellent eyesight, which of course he possessed, was critical to an observer's ability to discern features on the surface of Mars in telescopic images. The fact that he and Schiaparelli could see Martian structures that others could not see was, he believed, simply a physical failing on the parts of others.

On July 23, 1896, Lowell Observatory opened for business. The telescope itself was transported to Mexico and back again, over the next year, but soon found a permanent home on Mars Hill. Over the next two decades, up until the time of his death in 1916, Lowell would push his staff to undertake a systematic campaign to observe and study Mars.

Before he had built his observatory, let alone acquired any new data for Mars, Lowell made his views and his goals clear when he stepped into the public arena and gave a speech to the Boston Scientific Society in early 1894, the text of which was subsequently published in the *Boston Commonwealth* on May 26. Lowell explained his project in Arizona as one that would be "an investigation into the condition of life on other worlds . . . [for which] there is strong reason to believe we are on the eve of pretty definite discovery in the matter." Then, in discussing Schiaparelli's discoveries of canali on Mars, he told readers "the amazing blue network on Mars hints that one planet besides our own is actually inhabited now."[12] Lowell's observing program was designed to prove, rather than test, this claim.

Lowell's first observing campaign of Mars, conducted in 1894 with a 12-inch refracting telescope borrowed from Harvard and an 18-inch refractor lent to Lowell by Pittsburgh's Allegheny Observatory, was enormously successful. That first winter, Lowell Observatory astronomers observed 183 different canals, including virtually all the canals originally identified by Schiaparelli, and at least one hundred not previously seen by Schiaparelli. They found eight instances among these of canal doubling (gemination),[13] and some of the 183 canals were identified over and over again, on more than one hundred separate occasions. They also observed fifty-three different small dark spots first seen by William Pickering in 1892 and labeled by him as lakes.

In 1896, Lowell gained strong support for his work from the work of Leo Brenner (the pseudonym used by the Serbian rapscallion Spiridon Gopčević), who, in a communication published by the British Astronomical Association, reported that he had seen 102 canals, "70 of Schiaparelli, 12 of Lowell, and 20 new ones." Brenner also "saw [eight of] the Lowellian lakes" and even discovered a new one.[14] Brenner was a high school dropout, a liar, and a scoundrel. He had no credentials as an astronomer, but had the wealth to build Manora Observatory in Lussinpiccolo (now Mali Lošinj), on the island of Lošinj, off the coast of western Croatia. Almost certainly he made up everything he claimed to have observed and discovered, including the absurd rotation period of Venus of 23h 57m 36.27728s, but for at least a few years he was able to publish his reports in the professional journals *The Observatory, Journal of the British Astronomical Association*, and *Astronomische Nachrichten*, and his reports about Mars confirmed everything Lowell saw and claimed about Mars.[15]

Lowell's two-decades-long Mars observing campaign confirmed in his own mind his hypothesis, of which he was certain of the answer before he even began his study: Mars was inhabited by an intellectually advanced civilization. He began telling this story publicly during the winter of 1894–1895 in a six-article series he published in *Popular Astronomy*. He then authored a four-part series on the same topic that was published in *Atlantic Monthly* over the summer of 1895 and gave a series of public addresses that year on his Martian hypothesis. Finally, he published his first book on Mars, entitled simply *Mars*, in late 1895. Lowell clearly intended to persuade the public of his views. Then he expected the enthusiastic support of his worldwide popularity would convince the professional astronomers of the validity of his ideas.

Early on, Lowell occasionally drew mild but very cautious support from Princeton University professor of astronomy Charles Young, a well-respected scientist and the author of the most widely used astronomy textbooks of the era. Writing in the *Boston Herald* in October 1896, Young wrote carefully in detailing what was known about Mars without controversy and what was surmised about Mars with different degrees of uncertainty (How much water vapor is in the Martian atmosphere? What is the surface temperature of Mars? Are the dark lines on the surface canals?). He neither fully embraced nor rejected the canals hypothesis; he also speculated about life-forms on Mars, but remained noncommittal. In the end, he wrote eloquently and wisely that we should all "be cautious in accepting as ascertained truth the startling conclusions and unverified discoveries of imaginative observers. It is easy to see what one expects and wishes to find, especially on a disc so small and delicately marked as that of *Mars*."[16]

In the same month in which Professor Young was treating Lowell with professional courtesy, giving Lowell a small bit of the respect he desperately sought from the professional astronomy community, Lowell made publicly known, via a news release, that he had obtained new observational results about Venus that were a tour de force. Venus, he reported, rotates at exactly the same rate, 224.7 days, with which it orbits the Sun. Consequently, like Earth's Moon, Venus must always show the same face to the Sun. Lowell also proclaimed that he had discovered that "Venus is not cloud-covered, as had been supposed."[17] With these two important claimed discoveries about Venus, Lowell had apparently surpassed the achievements of the rest of the astronomy community. He was not done, though. He also published maps of Venus that included a series of dark lines, like spokes, that he asserted were permanent features of the Venusian surface.

Despite the public face of success, Lowell remained an outsider. While he had the financial resources to hire professional astronomers to work with him, he could not command their respect. Perhaps because of his own lack of professional experience and credentials, he quickly came to suspect many of the astronomers he hired to assist him at his observatory of disloyalty—both to him and his claims about Mars. Daniel A. Drew, hired in 1894 in time for Lowell's first Martian campaign, departed in June 1897. Wilbur A. Cogshall, hired with Drew, left in October of the same year. Samuel Boothryd, hired in 1897, abruptly departed in 1898. Thomas Jefferson Jackson See, an expert observer also hired for the 1894 campaign, and whose arrogance and egotistical nature generated no end of animosity toward him from his Lowell colleagues, was fired in July 1898. In 1901, Lowell discharged Andrew E. Douglas, who had joined the Lowell team in early 1894 to make atmospheric measurements to help determine the best location for Lowell's telescope, for "untrustworthiness." Douglas had complained to Lowell's brother-in-law, William L. Putnam, who was handling the finances for Lowell Observatory, that Lowell's "method is not the scientific method and much of what he has written has done him harm rather than good." His concluding advice was harsh: "I fear it will not be possible to turn him in to a scientific man." Not surprisingly, after Putman showed this letter to Lowell, Douglas found himself unemployed.[18]

To rebuild his staff, on the recommendation of none other than Cogshall, in 1901 Lowell had hired Vesto Slipher, who at the time was a graduate student at Indiana University. Slipher remained loyal to Lowell throughout Lowell's life and to Lowell's legacy afterward and spent his entire professional career in the employ of Lowell Observatory. In short order, Lowell also hired Carl Otto Lampland, who worked at Lowell Observatory from 1901 until 1951. Lampland distinguished himself by making great advances in astronomical photography, especially for observations of planets. A few years later, in 1906, Lowell filled out his professional staff for the long term when he hired Slipher's younger brother Earl, who also remained loyal to Lowell and Lowell Observatory, remaining there until 1964.[19]

As a result of Lowell's constant prodding, Vesto Slipher spent several years working on equipment and techniques that would allow him to measure the water vapor content of the Martian atmosphere. In 1908, when Slipher detected a stronger signature in the "a" line in his observations of Mars than in his "a" line observations of the Moon, Lowell trumpeted this achievement as definitive proof of the presence of water on Mars and as an implicit confirmation of all of his theories about Mars. With his ability to generate positive publicity for the work of his observatory, news of Slipher's discovery spread quickly.

Lowell's storytelling about Mars and Martian life peaked with his publication in 1906 of his book, *Mars and Its Canals*, in which he presented arguments in favor of the idea that life evolves naturally and inevitably from chemical processes. Consequently, the existence of living things on Mars, as well as elsewhere in the universe, is inevitable. He argued further that if plant life exists, then higher forms of life are also likely. As for Mars, since it has all the right ingredients for the chemical origin of life, the evolution of living things on that planet must have occurred.[20]

From his voluminous measurements of the canals, he determined that a "wave of darkening" swept from the poles to the equator at a speed of 51 miles per day. He concluded that this wave of darkening corresponded to the speed at which water was artificially pumped through the canals, thereby allowing vegetation to rapidly spread and bloom. Lowell believed that one of his strongest arguments in favor of this theory involved a multi-year study of a large dark region known (then and now) as Mare Erythraeum. He observed that the color of this region, located about 24 degrees south of the equator, changed from blue-green to chocolate brown as Martian winter set in; then Mare Erythraeum changed back to green with the onset of Martian spring.[21]

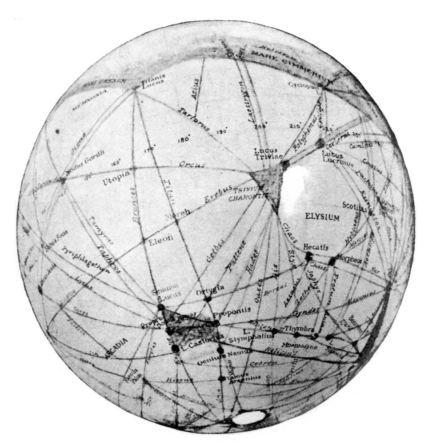

Figure 8.2. Percival Lowell's map of Mars, showing a globe-encircling canal system. According to Lowell, the canals were used by Martians to transport water from the water-rich polar caps to equatorial regions like the junction region near Elysium, identified as Trivium Charontis. Image from Lowell, *Mars as the Abode of Life*, 1908.

After more than a decade of writing and speaking and putting his observatory staff to work studying Mars, Lowell had achieved the impact he desired. He had transformed a debate among professional astronomers about the canals on Mars into an international, public debate about Martians. "Look back upon the year 1907 and pick out what has been, to your mind, the most extraordinary event of the twelve months," asked the anonymous author of a page 1 article in the *Wall Street Journal*, at the end of December. "Certainly it has not been the financial panic which is occupying our minds to the exclusion of most other thoughts." No, "The most extraordinary development has been the proof afforded by astronomical observations of the year that conscious, intelligent human life exists upon the planet Mars." The *Wall Street Journal* article reported, accurately, that "This proof is indeed circumstantial," but it has been sufficient to "produce circumstantial proof of the existence of intelligent life upon that globe." As for the significance of this discovery, "There could be no more wonderful achievement than this, to establish the fact of life upon another planet." All the credit, the writer noted, goes to Percival Lowell.[22]

To a great extent, Lowell's strategy had worked. He would publish one more book about Mars, *Mars as the Abode of Life*, in 1908. While working on that manuscript, Lowell was invited by the editor of the by-then-esteemed British journal *Nature* to present "an authoritative statement" about Mars and the Lowell Observatory observations of Mars that had been made in 1907. Lowell described the most recent observations made at Lowell Observatory: the melting of the southern polar caps, the "awaking activity" of the Lake of the Sun, the canals running north and running south and changing color, and the real-time visual evolution of the planet. "It is a direct *sequitur* from this that the planet is at present the abode of intelligent, constructive life," he wrote. More than that, "I may say in this connection that the theory of such life upon Mars was in no way an *a priori* hypothesis on my part, but the deduced outcome of observation, and that my observations since have fully confirmed it. No other supposition is consonant with all the facts observed here."[23] With supporting articles appearing in many major newspapers and affirmation by his invitation to write about Mars in *Nature*, the issue of canals and intelligent Martians appeared to be settled, with Lowell the intellectual victor.[24]

reality

Several measured aspects of the Martian environment that would be critical to Lowell's theory of life on Mars all seemed, surprisingly, to align in his favor. Life, Lowell believed, requires water, and water apparently had already been found on Mars. In addition, for life to exist on Mars, Mars must have a supportive atmosphere. Lowell Observatory astronomers worked hard to measure the thickness of the atmosphere on Mars, and they determined that the surface pressure was one-twelfth that of the surface pressure of Earth's atmosphere. While quite low, they deemed the Martian atmosphere more than sufficient to allow life to exist there. Lowell argued that the known existence of water vapor in the Martian atmosphere makes almost certain that other gases abundant in Earth's atmosphere should also be abundant in the atmosphere of Mars. Furthermore, Lowell marshaled circumstantial evidence in the form of the gravitational strength of Mars to support his hypothesis: the gravity of Mars is strong enough to prevent all gases except the lightest, hydrogen, from escaping into space. Therefore, oxygen and carbon dioxide could not escape into space and must be present on Mars at levels similar to terrestrial abundances of those important gases. Finally, in July 1907 Lowell produced a complicated calculation,[25] taking into account Martian cloud cover, reflectivity, heat absorption by the air and ground, and air pressure, from which he determined that the temperature at the surface of Mars was +48°F (+9°C). His answer enabled him to conclude that surface temperatures on Mars were similar enough to those on Earth to be compatible with life. (Lowell's method for calculating the surface temperature on Mars was challenged almost immediately. In December 1907, British physicist John Poynting published his own calculations,[26] taking into account what he dubbed the "greenhouse effect." His answer: between −15°F (−26°C) and −44°F (−42°C). Poynting was closer to correct than Lowell.)

Modern astronomical results have shown that all of these so-called truths about Mars are wrong. Mars is very dry, and the nineteenth-century measurements of water vapor in the atmosphere of Mars were all incorrect.

In addition, the atmosphere of Mars is more than fifteen times thinner than the Lowell Observatory staff thought. In an ingenious experiment, in July 1965 the Mariner 4 radio science team, including Gerald S. Levy, to whom this book is dedicated, transmitted radio signals from Mariner 4, when it was behind Mars, through the Martian atmosphere and onward to Earth.[27] In doing so, they showed how the atmospheric pressure, density, and temperature could be determined by measuring the changes in the properties of the radio signals. The answer from Mariner 4 was that the surface pressure on Mars is between four- and seven-thousandths of the atmospheric pressure at the surface of Earth (more recent measurements have converged on six-thousandths of one atmosphere).

The Mariner 4 science team also found that the surface temperature of Mars, which can vary with Martian latitude and Martian seasons, was between −154°F (−103°C) and −136°F (−93°C). More recent surface temperatures reveal a Martian low of about −243°F (−153°C) and a Martian high (near the equator in Martian summer) of about +68°F (+20°C). The average surface temperature on Mars is about −80°F (−62°C), making the surface typically more than 100 degrees colder than Lowell thought.

Quite importantly, Mars has no atmospheric oxygen. The Martian atmosphere consists almost entirely (96 percent) of carbon dioxide, which makes Mars extremely un-Earthlike.

The bottom line: Lowell was wrong about virtually every property of the Martian atmosphere he or other Lowell Observatory astronomers measured or calculated. Mars is an extreme version of a terrestrial desert: extraordinarily dry, with an enormous swing in temperature from day to night.

Lowell made the case for the artificiality of the canal system on the basis of the canals' straightness, the uniformity of their widths, and the fact that they were oriented like spokes on a wheel, with the central locations being those dark spots, which he described as oases with diameters of about 150 miles.[28] Lowell's efforts to convert both the public and the professional astronomy communities to his thesis that Mars was inhabited did draw fire from those who disagreed with him. His professional opponents focused their enmity on Lowell's claims for detecting the Martian canals as, without support for a Martian canal system, support for Lowell's intelligent Martians would also go away.

Among his most prominent opponents was W. W. Campbell, the director of Lick Observatory, who had the ability to muster the support of others in his battle against Lowell. Campbell organized the 1908 expedition of Lick Observatory staff to Mount Wilson to measure the water vapor content of the Martian atmosphere, which

 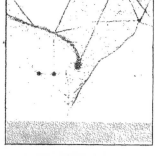

FIGURE 10.   October 5.   FIGURE 11.   November 3.
From a photograph.   Professor Lowell.
by Professor Hale.   24-inch Refractor.
60-inch Refractor.   stopped down to some 15 inches.
Views of Syrtis Major and Lacus Moeris 1n 1909 with various telescopes.

Figure 8.3. Comparison of a photograph of a portion of the surface of Mars (left), obtained in 1901 by George Ellery Hale using the then-biggest, and only recently dedicated, telescope in the world at Mount Wilson Observatory, above Los Angeles, with a hand-drawn sketch of Mars (right) obtained in the same year by Percival Lowell at Lowell Observatory in Arizona. The photographs show clearly that the straight lines identified on the surface of Mars by Schiaparelli and Lowell are actually broad and indistinct regions that reflect less light than other surface regions. Image from Antoniadi, *Popular Astronomy*, 1913.

refuted previous measurements that purportedly detected water on Mars. He also conducted a letter-writing campaign to four different journals, rebutting Lowell's claims that the site of Lowell Observatory allowed observers there to see through Earth's atmosphere with greater clarity than could astronomers at Lick Observatory or other major observatories. With Campbell's encouragement, George Ellery Hale, who had built the two largest telescopes in the world up until that time, the 40-inch refractor at Yerkes Observatory in Wisconsin and the 60-inch reflector at Mount Wilson in California, studied Mars in 1909 with the 60-inch. Hale reported to Campbell, "You may be surprised to know that they fail to show any trace of the canals!"[29] Without any doubt, Campbell was not at all surprised.

Campbell encouraged E. E. Barnard to use Hale's 60-inch telescope for his own study of Mars. Barnard was one of the most famous astronomers of that era; he was a pioneer of astronomical photography who discovered dozens of comets as well as the second closest star to Earth, known today as Barnard's star. Barnard had already waded into the debate over the Martian seas and canals through his observations of Mars in 1892 and 1894, made with the 36-inch telescope at Lick Observatory, shortly before he relocated to Yerkes. At that time, he published his conclusions in the *Monthly Notices of the Royal Astronomical Society*, saying that the "seas" and "oceans" of Mars looked, to him, like vast canyons and deserts, similar to much of California: "there was no suggestion that the view was one of far-away seas and oceans, but exactly the reverse." He also casually mentioned that "No straight hard sharp lines were seen on these surfaces, such as have been shown in the average drawings of recent years."[30]

In 1905, Barnard had visited Lowell Observatory to inspect the original photographs of Mars, taken that year by Lowell staff astronomer Lampland. Lampland had arrived at Lowell as a skilled astrophotographer. At that time, astrophotography of planets was considered far more difficult that taking photographs of stars or comets; many even considered photography of planets impossible. Motivated by Lowell's prodding, Lampland developed new techniques—specially designed, low-absorption glass for the 24-inch refracting telescope; observing only when the atmospheric conditions did not blur the images; carefully designed color filters to reduce the blurring that occurs when different colors of light come to a focus at different locations; a very steady "driving clock" (i.e., the clock that counters the rotation of Earth itself) that controlled the speed at which the telescope followed Mars across the sky. Lampland's efforts were major improvements that deservedly earned him the Royal Photographic Society medal in 1905 and that, beginning in May 1905, enabled him to take 700 photographs of Mars. Percival Lowell told newspapers throughout the United States and Europe about how sensational these photographs were. "These [photographic] plates, when sufficiently good, all show the canals," Lowell wrote in "First Photographs of the Canals of Mars," which he published in the *Proceedings of the Royal Society of London* in December 1906.[31] He published articles about these photographs in *Popular Astronomy*, *Astronomische Nachrichten*, the *Bulletin de la Société Belge d'Astronomie*, the *Harvard College Observatory Bulletin*, and of course his own *Lowell Observatory Bulletin*. The press covered these photographs as a major paradigm shift in the debate over the Martian canals. Lowell himself told the world about these photographic discoveries in his 1906 book, *Mars and Its Canals*. The photographs in this book confirmed to the world, he asserted, that Mars was a world dominated by Martians far more advanced than Earth's humans. In Lampland's photographs, "well-known features [of Mars] came out; 'continents' and 'seas', 'canals' and 'oases', the curious geography of the planet printed for the first time by itself in black and white." The photographs even captured "the first snowfall of the season, the beginning of the new polar cap . . . Upon the many images thirty-eight canals were counted in all, and one of them, the Nilokeras, double. Thus did the canals at last speak for their own reality themselves."[32] A correspondent for the London *Daily Graphic* agreed with Lowell, writing that these photographs of Mars "will put an end to the most hotly-debated controversy which in recent years has disturbed the lives of celestial observers."[33] A highly respected science writer employed by the Hearst newspaper chain agreed, after personally inspecting original photographic prints. "We may now look upon the existence of at least the principal Martian 'canals' as established."[34] Other newspapers echoed these sentiments, as did many professional astronomers, though few if any of them had seen the actual photographs, because mass-market publications found themselves unable to print reasonable facsimiles of the photographs. Additional photographs obtained by other Lowell staffers in South America during the 1907 Mars opposition stoked this fire higher when they were published in December and January issues of the magazines *Century* and *Cosmopolitan*.[35]

Barnard, on the other hand, after inspecting the original photographs, concluded that "they did not show the canals as claimed." Under pressure from Campbell, Barnard conducted his own observing campaign to observe Mars in late 1910, from which he again found no evidence for canals.[36]

Another Lowell opponent was Simon Newcomb, one of the most distinguished mathematicians of the nineteenth century. He was a member of the National Academy of Sciences and a fellow of the Royal Society. He had served as president of the American Mathematical Society and as the first president of the American Astronomical Society. His list of awards (Gold Medal of the Royal Astronomical Society, Copley Medal of the Royal Society, Chevalier of the Légion d'Honneur) was lengthy. Having Newcomb on your side in an intellectual skirmish was much preferable to taking him on as an opponent, but he opposed Lowell and Lowell's canals. Newcomb staked out his position as early as 1897, writing:

> While every astronomer has entertained the highest admiration for the energy and enthusiasm shown by Mr. Percival Lowell ... they cannot lose sight of the fact that the ablest and most experienced observers are liable to error when they attempt to delineate the features of a body 50 to 100 million miles away through such a disturbing medium as our atmosphere.... That certain markings to which Schiaparelli gave the name of canals exist, few will question. But it may be questioned whether these markings are the fine sharp lines found on Schiaparelli's map and delineated in Mr. Lowell's beautiful book.[37]

Newcomb was persistent in publicly questioning Lowell's ability to see the details on the Martian surface that Lowell, on the basis of the self-proclaimed excellence of the Flagstaff observing location and of his own eyesight, claimed the ability to observe. After carrying out a series of tests on the ability of the human eye to discern black lines against a lighter background at a distance of about 100 feet, Newcomb concluded that Lowell's measurements were impossible. He nevertheless allowed that although "these results may weaken the probability of the reality of the entire canal system, it does not disprove its possibility."[38]

Significant problems arose with all of these new discoveries made by Lowell. No other astronomer could see the surface markings on Venus, and, to this day, no other astronomer has ever claimed to have seen the surface of Venus using a telescope on Earth. We now know, in fact, that an enormously thick atmosphere shrouds Venus, and Lowell could not possibly have seen any permanent surface features on Venus. As for the rotation period of Venus, no other astronomer was able to confirm Lowell's measurement, and we now know that it too is wrong. Venus does spin slowly, rotating once every 243 days, but it does so in the opposite direction than that measured by Lowell. With the publication of his maps of Venus in December 1896, Lowell turned himself into a marked man among those whom he thought of as professional colleagues. They did not think of Lowell, who had no professional training or credentials as an astronomer, as a professional equal, and his studies of Venus turned him into a pariah among astronomers.

One of Lowell's early supporters, Eugène Michael Antoniadi, metamorphosed into one of his most vocal opponents. Antoniadi, who was born and trained as an astronomer in Constantinople (now Istanbul), began his career by accepting a position in 1893 in France working with Flammarion at Meudon Observatory, at the Observatoire de Paris. While there, he was personally able to observe, he thought, more than forty Martian canals. After being named the new leader of the Mars section of the British Astronomical Association (BAA), he slowly began to see Mars differently. In his new leadership role, he collected the many observations of Mars made by Association members and from them published annual reports. In his first reports, he remained an enthusiastic supporter of the canals hypothesis, as well as of Flammarion's and Lowell's claims for life on Mars. His report for 1896–1897 stated that "the canals were seen by all the working members of the section invariably." He also reported that dark areas were "very likely lands covered with scanty [*sic*] *grassy* vegetation," rather than what another member of the BAA had called "ruddy vegetation." Continuing, he noted that "The dark areas possibly represent both vegetation and water. Vegetation where seasonal changes are recorded; water where general invariability of the tint is the predominant feature."

In an early dose of skepticism, however, Antoniadi declared that another BAA observer "was certainly right in attributing the doubling of the linear features to the secondary images formed in the eye under imperfect vision." In fact, Antoniadi had begun performing experiments in which he proved to his own satisfaction that single lines blurred into double lines as a result of poor vision. The "artificial doublings," as he now referred to these phenomena, were best attributed to a "focusing error." The doublings were not real, even if the single canals, themselves, were.[39]

Two other astronomers were able to see more canals than had Antoniadi, but he credited himself with the sighting of forty-six canals. Antoniadi also commented that seeing the canals was not easy. If he did not already know of their existence, from the work of Schiaparelli, he said that he would have missed seeing most of them. As a result, Antoniadi was cautious and excluded from his summary the work of one C. Roberts, who had recorded at least 134 canals in his drawings, using only a 6.5-inch telescope. After the 1898–1899 observing campaign of Mars, he issued another report, in 1901, in which he wrote, "Notwithstanding the natural skepticism of many scientific men, every opposition brings with it its own contingent of confirmations of Schiaparelli's discovery of linear markings, apparently furrowing the surface of Mars."[40]

Antoniadi's cautious skepticism was growing, however. He took note of "Mr. Maunder's and Mr. B. W. Lane's valuable experiments, which show the 'canaliform illusion' to be a physiological phenomenon of at least some eyes." He noted, further, that "one-half of the 'canals' seen by observers are the boundaries of faint half-tones."[41] As a result, he began to wonder about the ability of the human eye and brain to create imaginary structures from blurry images. In addition, as Lowell's observing team started to report seeing linear markings on the planets Mercury and Venus, as well as on some of the large moons of Jupiter, Antoniadi wondered further whether all of the lines on all of the planets were optical illusions or real features demarking regions that reflected different percentages of sunlight (in the language of planetary astronomy, regions with different albedos). Antoniadi's 1898–1899 report included a new explanation for at least some of the canals: "we know now for certain that a considerable number of so-called 'canals' coincide with boundaries of adjacent areas with different albedos. The fact is significant." In other words, some features identified as canals were simply the borders between two areas of the surface with different colors that reflected different percentages of sunlight. In his view, the evidence for canals was fading. Antoniadi also reported that "'Lakes' forming on the intersection of widening canals are not necessarily real, but might be superposition images."[42]

By 1902, he was no longer a fervent supporter of the Martian canals hypothesis, though he had not yet become an opponent, either. He also was no longer working with Flammarion. His report on the 1903 Mars opposition campaign marked a major change, in that he published two charts of Mars, one with and one without canals.

He then temporarily escaped the Mars debate by spending half a decade studying architecture. But in 1909, when the director of Meudon Observatory offered him use of the Grand Lunette, the largest telescope (33-inch diameter) in Europe, for the study of Mars, he embraced the opportunity.

By all accounts, the night of September 20, 1909, was historically one of the best nights, if not the single best night ever, for anyone observing Mars with a telescope from Earth before the modern age. Because of an apparent temperature inversion over Paris, Antoniadi had clear, still air above the Grand Lunette for seven hours, offering him the best-ever images of the red planet obtained by an astronomer up until that time, and ones that would not be surpassed until the age of orbiting telescopes and interplanetary spacecraft. And what did Antoniadi see? No canals. Not even any straight lines. Nothing that anyone could identify with the imaginative Martian drawings of Lowell and Schiaparelli. He reported that "The planet appeared covered with a vast and incredible amount of detail held steadily, all natural and logical, irregular and chequered, from which geometry was conspicuous by its complete absence."[43] By the time Antoniadi submitted his *Sixth Interim Report for 1909, dealing with some further Notes on the so-called "Canals,"* he had taken firm charge of the anti-canals movement. In presenting the report of one observer who had seen a handful of double canals, Antoniadi asserted without equivocation, "It is high time that evidence of this kind should be pronounced worthless; and but for such uncertain data, usually obtained with inferior telescopes, there never would have been a question of canals on Mars." He continued by remarking, "However welcome such phenomena may be to the believer in canals, they can have really no other meaning than that of showing a very complex and irregular structure under a fallacious geometric form." As for Schiaparelli's canals, "A fact of the highest import is that, under good seeing, almost all of Schiaparelli's 'canals' visible either broke up into most complicated and irregular groups of shadings or became the mere indented edges of these shadings." Antoniadi wrapped up his remarkable takedown of the Martian canal hypothesis with this bold conclusion: "We have, of course, no more right to speak of the true canals on Mars than of the dykes of Mars or the roads of Mars. Whether such things exist or not on the planet we cannot know; and any consideration regarding them must be treated as unwarranted speculation. The term *canal* has no more relevancy on Mars than *sea* on the Moon."

An addendum to this summary included a report on observations made by George Hale with the "most powerful instrument in the world," the 60-inch telescope on Mount Wilson. Hale reported, "no trace of narrow lines, or geometric structure, was observed. A few of the larger 'canals' of Schiaparelli were seen, but these neither narrow nor straight." Antoniadi had the final word: "It is hoped that the above letter will dispel the last doubts to the true character of the markings on the planet. The frail testimony of small refractors has vanished before the decisive evidence of giant instruments; and the telescopes of Princeton, Lick, Yerkes, Mount Wilson, and Meudon have settled the question forever."[44]

Antoniadi completed his work of finishing off the canals of Mars by taking his conclusions directly to Lowell's audience of lay readers. Writing in *Popular Astronomy* in 1913, he told readers, "Ponderous volumes will still be written to record the discovery of new canals. But the astronomer of the future will sneer at these wonders; and the canal fallacy, after retarding progress for a third of a century, is doomed to be relegated into myths of the past."[45] Antoniadi was right and prescient. Lowell's work did retard scientific progress in planetary astronomy for most of Lowell's lifetime; worse still, Lowell's work also gave planetary science a bad name for most of the first half of the twentieth century. The rebirth of planetary astronomy as a significant scientific discipline emerged only after the launch of Sputnik, with the invention of interplanetary spacecraft.

The next time Mars was positioned well for viewing, in 1924, University of California astronomer Robert Trumpler made use of the fourth largest telescope in the world at that time,* the 36-inch refractor at Lick Observatory, to carry out "a careful study of the surface of Mars by photographic and direct visual observation." He obtained about 1,700 photographs of Mars. He was able to see "over thirty of

*The largest refracting telescope was the 40-inch at Yerkes Observatory, Wisconsin, completed in 1895. The 60-inch (completed 1908) and 100-inch (completed 1917) reflecting telescopes on Mount Wilson were also larger.

the canals," though "the so-called canals were not found to be the sharp fine lines as which they have been sometimes described. Even under the best atmospheric conditions the canals appeared somewhat diffuse and the narrowest ones not less than 25 miles wide." With a mild dose of skepticism, Trumpler concluded, "it seems that under the term canal a variety of markings are included sometimes quite dissimilar in character."[46] Trumpler implies a naturalness to these features, but nevertheless was trapped by Lowell's legacy and vocabulary into discussing oases and canals. Trumpler would go on to have an extremely distinguished career—in 1930, he made one of the most important astronomical discoveries of the twentieth century, proving the existence of small grains of dust in interstellar space, between the stars, which absorb light from distant stars and make them appear farther away than they actually are. His study of Mars, however, was not his best moment, and it gave one last tiny breath of life to Lowell's Martian canals hypothesis.

Antoniadi, for all intents and purposes, brought down the curtain on Lowell's canals and intelligent Martians when he published his book *La Planète Mars* in 1931. With an entire chapter devoted to "L'illusion des canaux," he banished Martian life, at least in the form imagined by Percival Lowell, to the ash heap.

chlorophyll, lichens, and algae

chlorophyll

While Percival Lowell's ideas about canals and life on Mars may have been losing favor among professional astronomers in the west, his influence nevertheless continued to extend eastward into Russia, where they guided the work of astronomer Gavriil Adrianovich Tikhov. Tikhov decided to prove that vegetation existed on Mars by looking for evidence of chlorophyll in the colors of Martian light. Born in Minsk in 1875, Tikhov trained first at Moscow University and then continued his education at the Sorbonne in Paris. He worked as an astronomer at Pulkovo Observatory for almost four decades. He served as a pilot watcher in the First World War and survived that war; he then endured and survived both the Russian Revolution and the civil wars that followed, always continuing his astronomical work.[1]

Tikhov began his searches for chlorophyll as early as 1909. He knew that light reflected from chlorophyll-containing vegetation on Earth appears green, so presumably he was looking for clues, using homemade colored filters, that Mars had green surface patches. Most plants on Earth do look green and, indeed, chlorophyll is the reason. The chlorophyll molecules (there are two different kinds of chlorophyll in plants, chlorophyll a and chlorophyll b) power photosynthesis; they do so by absorbing sunlight and transferring that solar energy to an electron that kick-starts the process of converting water and carbon dioxide into sugar and oxygen. Chlorophyll molecules are very good at their job of absorbing the energy in sunlight. Together, chlorophylls a and b absorb nearly 50–90 percent of the light in the violet and blue and 50–60 percent of the photons in the red. But they are poor absorbers—and therefore great reflectors—in the green and yellow.

Tikhov was unable to find any evidence for green reflected light in his earliest observations for chlorophyll on Mars. Yes, Mars has dark patches, but they simply don't look green! Undaunted, and apparently unaffected by the evidence that seemed to have proven his initial hypothesis wrong, he planned to continue his search for chlorophyll on Mars during the next set of favorable Martian oppositions in 1918 and 1920. He designed and fabricated a primitive spectrograph for this purpose and didn't let World War I or the Russian Civil War deter him. Despite the battles that took place around the Pulkovo Observatory during these years, he observed Mars, as planned; but he was unable to uncover evidence for chlorophyll in his Martian spectra. Two decades later, during the siege of Leningrad (formerly known as Saint Petersburg), which commenced in September 1941, Pulkovo Observatory was destroyed. After the war, Tikhov relocated to Kazakhstan and restarted his astronomical work at Alma-Ata Observatory.

As has often been stated, the absence of evidence is not evidence of absence, and in this case the absence of a definitive detection of chlorophyll on Mars did not, in his view, demonstrate that plant life did not exist, somewhere, on Mars. Imagination and belief trumped observational evidence; he allowed himself to believe that his measurements simply demonstrated that plant life on Mars must be different from plant life on Earth and that his telescopic evidence revealed that vegetation on Mars must grow without the benefit or need of chlorophyll. Undaunted by his apparent observational proofs that Martian life lacked chlorophyll, while at Alma-Ata he investigated the colors of terrestrial vegetation as a means toward gaining a better understanding of the colors of Mars. This line of reasoning led Tikhov to invent the field of research he called astrobotany, in which he would study the reflected light from plants, in particular plants that grew in extreme, Mars-like environments on Earth—at high elevation or in extremely low temperature environments—in order to look for spectra from plants that might lack the green-color signature of chlorophyll. Amazingly, he found that at very low temperatures the color of the reflected light from chlorophyll was not always green for some plants. In addition, some plants, when grown at low temperatures, will have colors other than green.

The robustness of conclusions with regard to the presence of vegetation on Mars depended strongly on the likelihood that the surface temperature of Mars was conducive to plant life, but astronomers in the nineteenth century had only been able to guess at that temperature, and the temperature calculations made by Lowell and Poynting in 1907 were neither accurate nor definitive.

In the early 1920s, William Weber Coblentz, a physicist who spent his entire career working for the National Bureau of Standards in Washington, DC, decided to measure the temperature of Mars. Working with Lampland at Lowell Observatory, he conducted a series of carefully planned observations to measure the intensity of light from Mars in the mid-infrared, at wavelengths from 8 to 15 microns. Using these measurements, and assumptions about the reflectivity of Mars and the effects of Earth's atmosphere on these measurements, Coblentz measured the temperature of Mars in different Martian seasons and across separate surface regions that he characterized as either bright or dark. He found that the bright areas, with temperatures as low as freezing (0°C or 32°F) are cooler than the dark areas, which had temperatures of 10–16°C (50–60°F). Then, having associated bright areas on Earth with hot deserts, he concluded that this bright-and-cold versus dark-and-hot bimodality for landforms on Mars "is just the reverse of conditions here on Earth, where the surface of the bare desert areas becomes burning hot."[2] His logic defied reality. The opposite conclusion—that Mars is similar in this respect to Earth—is also easy to reach: the brightest areas on Earth are the cold polar caps and the warmest areas are continental regions far from the polar caps, just as they are on Mars.

Coblentz drew additional significant conclusions about Martian vegetation based on these temperature measurements, given other

known conditions of Mars—most notably the relative dryness of
Mars in comparison to Earth. "The observed high local temperatures of Mars," he wrote, "can be explained best by the presence of
vegetation which grows in the form of tussocks or thick tufts, such
as pampas grasses, and the mosses and lichens that grow in the dry
tundras of Siberia." Coblentz obtained the answer he expected, the
answer he perhaps wanted, and he was partly right: the highest temperatures on Mars would allow for the growth of vegetation. Most of
the time on most of Mars, the temperature is close to $-16\,^\circ$C ($3\,^\circ$F) in
the daytime, with nighttime temperatures plunging down to $-90\,^\circ$C
($-130\,^\circ$F); however, noontime temperatures at the equator in the
summer can reach as high as $20\,^\circ$C ($68\,^\circ$F). He, of course, had absolutely no evidence that pampas grasses, mosses, or lichens actually
were growing on Mars, let alone that the Martian vegetation affects
Martian temperatures.

In the 1920s and 1930s, American astronomers Vesto Slipher and
Robert Trumpler and then Canadian astronomer Peter Millman independently pursued the same goal as Tikhov. Almost certainly, they
observed Mars in complete ignorance of the work of Tikhov, who
worked largely in isolation from western astronomers.

Slipher's 1908 studies of Mars were of marginal value, at best,
though Percival Lowell had trumpeted them as definitive evidence
for the presence of water in the atmosphere of Mars.

What research was Slipher carrying out in the 1920s regarding
Mars? Lowell Observatory, of course, by then had a long history of
observing Mars, and Slipher and his brother Earl had continued that
program of observations with regularity. Because Mars is well positioned for study from telescopes on Earth about every two years, not
surprisingly the publication cycle for the Slipher brothers' studies of
Mars followed this same cycle. One or the other of them published at
least four papers concerning observations of Mars in 1922, five more
in 1924, and four more in 1927.

Vesto Slipher began chasing down evidence of chlorophyll at the
urging of Percival Lowell during the years from 1905 through 1907,
with no success. Two decades later, the technology for taking astronomical photographs had improved. Slipher now had available for
his work new sensitizing dyes—pinaverdol for improving sensitivity

in the green and yellow, pinacyanol for the red, and dicyanin A and kryptocyanine for the infrared—that he used in darkroom baths to modify the emulsions on commercial photographic plates before use. These new chemicals were supposed to "bring out strikingly the red color of Mars," especially in comparison to the Moon. The reflection spectrum of chlorophyll, he noted, is bright in "the deep red, beyond the sensitivity of the eye." He decided to look for this "deep red" signature of chlorophyll because proof of the existence of chlorophyll on Mars would prove that life exists on Mars and would therefore add to the legacy of Percival Lowell. Apparently, his experimental results were negative for the presence of chlorophyll on Mars, though he would describe those results rather cautiously. "The Martian spectra of the dark regions," he reported in a paper published in 1924 in the journal *Astronomical Society of the Pacific*, "so far do not give any certain evidence of the typical reflection spectrum of chlorophyll."[3] In the years that followed, he was busy running Lowell Observatory and the search for Lowell's Planet X, studying the atmospheres of Venus and the giant planets, and obtaining spectra of faint extended nebulae (some of these being clouds in the Milky Way galaxy, others in distant galaxies), but he rarely observed Mars again and never again published any additional details on this particular project. Whether he simply lost interest in Mars or chose not to pursue a line of research that might prove embarrassing to himself or to the legacy of the man whose name graced the observatory he directed, we don't know.

By 1924, Trumpler's reputation was secure, in large part because of his role in carrying out observational tests in Australia on September 21, 1922, of Einstein's theory of relativity during a total eclipse of the Sun. While Sir Arthur Eddington had made Einstein world-famous when, during his eclipse expedition of May 29, 1919, he had photographed the bending of starlight by the gravitational pull of the Sun, Eddington's measurements were on the edge of what could be measured. In contrast, Trumpler's measurements of the deflection of starlight around the limb of the Sun represent the first secure confirmation of Einstein's theory of relativity.

A few years later, Trumpler discovered that all of space within the Milky Way is filled with a haze caused by interstellar dust, a discovery that would lead authors to write his name into every textbook published in astronomy since the 1930s. This haze dims the light of distant stars and has the effect of blocking out more blue light than red light, making distant stars look redder than their true colors. For his important contributions to astronomy, Trumpler was elected to the National Academy of Sciences in 1932.

Trumpler described his studies of Mars in the *Science News-Letter* in 1927, and though he dismissed the artificiality of the network of lines seen on the surface of Mars, he did identify a "close relationship between the network and the extended dark areas of Mars which are of a bluish-green tinge." This relationship, he wrote, "suggests the hypothesis that both are made visible by vegetation and that the network-lines represent lanes of greatest fertility."[4]

Trumpler's work influenced the opinion of none other than Princeton professor of astronomy Henry Norris Russell, whose work in creating the concept of what is now called the Hertzsprung-Russell diagram cannot be overstated in importance for all of astronomy. In 1926, Russell responded to an inquiry from a writer for the *Science News-Letter* about the recent observations made of Mars by noting that Mars has all the necessary conditions for life as we know it; in addition, he commented that the large green areas on Mars that change color with the Martian seasonal cycle make it probable that vegetable life exists on Mars.[5]

Fifteen years later, during the next observing period when Mars was well placed for study, Millman took up the challenge of looking for chlorophyll on Mars. Millman had cut his teeth, professionally, on Mars, as his first professional paper was a study of Mars, based on observations he had made while in Japan as a high school student at the Canadian Academy in Kobe. After concluding his studies at the University of Toronto and then spending four years at Harvard, Millman had attained the status of a specialist in the study of meteors.

Prior to 1939, nothing in Millman's career led him down the path to a study of vegetation and a search for chlorophyll on Mars. He had studied binary stars and discovered a Cepheid star in the constellation Scorpio. Cepheids are arguably the most important kind of star known to astronomers. In the first decade of the twentieth century, Henrietta Leavitt, working at the Harvard College Observatory, had discovered that Cepheids, which were known to cycle in brightness from bright to faint to bright again, did so in a very dependable way. The brightest Cepheids required more than one hundred days to complete one cycle, while the faintest Cepheids completed one cycle of brightness changes in only a day. Once this understanding of the behavior of Cepheids was quantified, astronomers found that they could measure the period of variability of the light for any selected Cepheid and use that information to determine the distance to that Cepheid and the cluster of stars in which that particular Cepheid is embedded. Edwin Hubble had used Cepheids to prove, in 1925, that the Milky Way was not the only galaxy in the universe. He then used them again in 1929 and 1931 to discover that the universe is expanding. In the late 1930s, Cepheids were still among the most important objects of study for astronomers, and in fact remain similarly important well into the twenty-first century. The discovery of a Cepheid was an important discovery in the 1930s and indicated Millman had strong observing skills, because the discovery of such an object requires very careful measurements made over many months or even years. He also had published numerous articles on how to photograph and make spectroscopic observations of meteors, fireballs, and shooting stars. Added to this, he wrote papers in which he analyzed the spectra of meteors and studied the frequencies at which meteors struck Earth. Then, sticking out of his list of professional publications like a green patch on Mars is his paper "Is there vegetation on Mars," which he published in 1939 in the journal *The Sky*, a magazine that, after only a few years of publication, was subsumed into *Sky and Telescope*.

Millman's work on this project represents some of the most rational, sensible scientific effort ever expended on the subject of life on Mars, and merits great respect for the care and caution he brought to the project. In his paper, he wrote, "So much nonsense has been written about the planet . . . that it is easy to forget that Mars is still an object of serious scientific investigation." He reminded his readers of the simple but compelling reasons many astronomers felt led to the hypothesis that Mars had vegetation.

Mars was well known to have polar caps that waxed and waned with the Martian seasons. When the southern polar cap shrank, presumably because the water ice it contained melted, "a wave of darkening proceeded from the polar regions towards the equator, and the seas, which in the winter were faded and inconspicuous, now became darker and greenish, starting with those nearest the pole. . . . It has been a natural and popular explanation that the seas are in reality vegetation which is nourished by the melting polar snow and goes through a seasonal cycle similar to the vegetation on earth. The greenish color of the seas has been considered an additional support to this hypothesis."[6] He then noted that this hypothesis should be tested, and he proceeded to design such a test. His approach was rational, logical, and unemotional.

In writing for his audience in *The Sky*, he explained that leaves are green because chlorophyll strongly reflects "yellow-green and yellow rays" but only weakly reflects light of shorter (violet, blue, blue-green) or longer (red) wavelengths. Chlorophyll, he continued, is also an extremely good reflector in the infrared. Thus, one could detect the signature of chlorophyll by looking for reflected light from Mars that was similar to reflected light from chlorophyll: strong green light, weak colors in the rest of the visible spectrum, and strong infrared light, just beyond the red. This approach, which was very similar to that used by Slipher a decade earlier, was the basis for Millman's observing strategy.

Millman designed an observing plan using the 74-inch telescope (which saw first light in 1935) of the David Dunlap Observatory, located near Toronto. First, he obtained photographs of Mars in which he isolated two dark-colored neighboring regions of Mars known as Syrtis Major and Mare Tyrrhenum.* Next, he photographed an adjacent, lighter colored region. According to the hypothesis he was testing, Syrtis Major and Mare Tyrrhenum were dark because they were covered with vegetation, while the second region was lighter in color because it was free of vegetation. A comparison between the

---

*In modern usage, all names for craters, mountains, valleys, and all other surface features on Mars must be approved by the International Astronomical Union's Working Group for Planetary System Nomenclature. The IAU first approved names for 126 surface features on Mars in 1958 (including Syrtis Major and Mare Tyrrhenum), approved three more names in 1967, 273 names in 1973, 528 names between 1976 and 1979, and has been kept busy approving names for Mars surface features ever since. See: https://planetarynames.wr.usgs.gov/Page/MARS/target

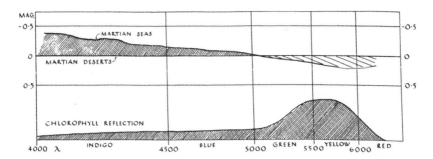

Figure 9.1. Plot of the intensity of reflected light from Mars (in units of stellar magnitudes) versus wavelength (in units of angstroms; 1 angstrom = one ten-billionth of a meter) or color of light. The upper part of the figure shows the intensity of reflected light from Martian seas (top line) and Martian deserts (bottom line). The seas are much brighter than the deserts in the colors indigo and blue, but are darker than the deserts in green and yellow. The bottom part of the figure shows the intensity of reflected light from chlorophyll, which is extremely reflective in the green and yellow and much less reflective at other colors. Clearly, neither the Martian seas nor deserts reflect light in a way that would be similar to the spectral reflectance signature of chlorophyll. Image from Millman, *The Sky*, 1939.

reflected colors of the two regions should show a distinct contrast, under the assumption that chlorophyll is present in Martian vegetation. Millman found, in his measurements, that although the dark colored seas "appear greenish in color to the eye," the fact that they were "relatively strong in violet, blue, and blue-green light but weak in yellow, orange, and red ... negates the existence of chlorophyll as an agent in producing that green color." Were chlorophyll present, the yellow-green and yellow light would have been more intense than the other colors.

To check for consistency, Millman performed the same experiment with genuine, Earth-grown, Canadian green leaves and obtained exactly the result he expected for materials containing chlorophyll. The Canadian leaves reflected light very strongly in the yellow-green and yellow colors relative to the other colors.

Millman drew a rational conclusion from his measurements. His results, he wrote, "do not give any definitive evidence about the existence or non-existence of vegetation on the Martian surface. What they do seem to indicate is that the greenish color of the seas cannot be taken as a support for the vegetation hypothesis, since the green color does not seem to be like the green reflected from our terrestrial leaves." He continued, again writing maybe all too reasonably, "Perhaps, after all, we are rather presumptuous to think that anything on Mars which might have some remote similarity to biologic life on the earth would develop under the same organic system as that found here." Millman never returned to this project and never again studied Mars.

lichens

By the 1940s, the idea that we shared our solar system with walking, talking Martians had faded away; however, the possibility that Mars harbored some kind of life remained strong. Given the technological advances made by scientists and engineers during World War II in research areas that could impact astronomy, it was only a matter of time before a bright, young astronomer used those new tools to discover new ways to study Mars.

Gerard P. Kuiper, one of the giants of twentieth-century astronomy, was that bright young astronomer. During the late 1940s and 1950s, Kuiper worked at the University of Chicago's Yerkes Observatory, in Williams Bay, Wisconsin, and at the University of Texas at Austin's McDonald Observatory, near Fort Davis in western Texas. In 1947, Kuiper used a then-new device, a lead-sulfide photoconductive cell developed for military purposes earlier in the decade, in a new detector system for the 82-inch telescope at McDonald Observatory. This device was a very sensitive infrared spectrometer, fully a thousand times more sensitive in the infrared than the previous generation of instruments. With this new detector system, Kuiper was inventing modern infrared astronomy and using his new tool to study the planets.

In 1944, in a paper written while on a leave of absence from the University of Chicago and Yerkes Observatories to conduct "war research" at the Radiation Laboratory in Cambridge, Massachusetts, Kuiper discovered methane gas in the atmosphere of Saturn's largest moon, Titan (Kuiper claimed that methane was the dominant gas in Titan's atmosphere, though we now know that nitrogen is far more abundant), and in 1948 he discovered Miranda, the fifth discovered of the moons orbiting Uranus. In 1951, using known information for the orbits of comets, he predicted the existence of a disk of objects in the outer solar system located more distant from the Sun than even Neptune. That band of the solar system, which was not found observationally until the 1990s, is now known as the Kuiper Belt*; and NASA's Kuiper Airborne Observatory, the KAO, was named to honor Kuiper, who died in 1973 after pioneering airborne infrared astronomy in the 1960s. The KAO was a C-141 military cargo plane with an infrared telescope built into the cargo hold. The KAO, which flew at altitudes as high as 45,000 feet in order to fly above 99 percent of the water vapor in Earth's atmosphere, operated from 1975 until 1995. Astronomers flying on and working in the KAO received high-altitude survival training—if the tiny control room depressurized, they had 15 seconds to safely secure their oxygen masks before asphyxiating.

Using the 82-inch telescope at McDonald Observatory in 1948 (the second largest telescope in the world when it was completed in 1938), Kuiper obtained the first-ever color picture of Mars, which was such an exciting achievement that, in June, *Life* magazine published his spectacular photograph. The headline of the article that accompanied this picture was rather mundane but very much in tune with mainstream astronomy at the time: "Newest photographs and studies disclose that life on planet is limited to simple vegetation." The text reported that the "green patches are vegetation, but of the lowest order, lichens, which live by drawing moisture from the air." Kuiper informed readers of *Life* that "the lowly lichens represent life's last stand on the Red Planet."[7]

One might surmise that Kuiper's photograph revealed something incredible about Mars: life. The remarkable news, however, was not that Kuiper had discovered life on Mars. The idea of life on Mars was old news; life on Mars was expected and assumed, known even, except for the details. No, the amazing news was that Kuiper thought

*In 1949, Irishman Kenneth Essex Edgeworth had made a similar prediction, suggesting the possibility of a stable reservoir of comets and small planets in the trans-Neptune zone of the solar system.

he had proven that life on Mars was limited to lichens. Not little green men, not forests of trees, not even algae or moss. Just rootless, stemless, leafless lichens.

The lowly and primitive lichen is actually two different organisms, living together in a symbiotic relationship. Most of the lichen is composed of long cells of fungi, connected end-to-end to form long, tubular filaments. Unlike normal plant cells, many fungal cells contain multiple nuclei; also unlike normal plant cells, the cell walls of fungi are strong, as they receive structural support from the presence of chitin, a carbohydrate polymer molecule. The other living part of a lichen, which lives among the fungi cells, is photosynthetic, usually a type of algae known as green algae but sometimes an ancient form of bacteria called cyanobacteria (which are often called blue-green algae, although they are not plants).

During the Mars opposition campaign that begin in late 1947 and extended into early 1948, Kuiper succeeded in definitively demonstrating that the atmosphere of Mars contained a small amount of carbon dioxide and that CO_2 was the dominant gas in the Martian atmosphere. Figuring out exactly how much CO_2 was present was a great deal more difficult, however, than demonstrating that it was the most important gas in the Martian atmosphere. The answer he obtained from his calculations for the amount of CO_2 in the Martian atmosphere was a bit too low, but he did correctly deduce that Mars's atmosphere was very thin in comparison to Earth's atmosphere.

As *Time* magazine noted in a report on Kuiper's discoveries published in March 1948, Lowell had "believed that life on Mars is older and more highly developed than it is on Earth. . . . It is at least possible [according to Lowell] that Martians reached a stage of scientific civilization while man's ancestors were still fish or reptiles. Perhaps the Martians have evolved by now into some superscientific stage."[8] Lowell's views of Mars and Martians, however, were outdated by 1948, and Kuiper's own discoveries about the presence and abundance of CO_2 on Mars revealed that Mars wasn't quite as hospitable as Percival Lowell had thought. The *Time* article on Kuiper's new research noted that "Last autumn he found that the atmosphere contains a small amount of carbon dioxide, which is necessary to plants (the basic living organisms). Without any carbon dioxide, plants cannot live, but too much would indicate that there are no plants on Mars to consume it."

What is missing from this explanation is that the atmosphere of Mars contains almost nothing except carbon dioxide. The Martian atmosphere does contain a small total amount of carbon dioxide, but the atmosphere contains almost nothing else. Kuiper suspected this was the case, though definitive proof of the thinness of the Martian atmosphere would not be found until two decades later. Kuiper did, nevertheless, assert that the climate of Mars resembles "earth at an elevation of 50,000 feet.... This probably would support lichen, since these plants act like sponges and suck up water vapor present in air. Rain is not necessary for their existence."[9] Kuiper's Mars was a cold, dry, nearly airless place. For anyone expecting astronomers to find Martians, Kuiper's news would have been disappointing.

Having figured this out mostly correctly, he moved on to study the polar caps of Mars. Solid CO_2 (dry ice), he decided, was unlikely at the relatively warm (for dry ice) surface temperatures of Mars and at the atmospheric pressure he had measured for Mars. But the absence of solid CO_2 was mere speculation. With his new infrared spectrometer, he could test his hypothesis, and so he set out to do so. "The observations of the Mars polar cap showed that the cap resembles a water-cell spectrum," he concluded. After repeated observations and laboratory tests to obtain spectral comparisons of CO_2 snow and terrestrial snow and frost, Kuiper had an answer: "The conclusion is that the Mars polar caps are not composed of CO_2 and are almost certainly composed of H_2O frost at low temperature." Again, *Time* publicly praised the genius of Kuiper: "Last week Kuiper focused his spectrometer on the gleaming ice cap, dwindling fast in the Martian May. It turned out to be 'water in the solid state.'"

In fact, Kuiper's solution to the polar cap problem was partly right and partly wrong. We know now that the polar caps have permanent caps of water ice overlaid by seasonal (winter) caps of carbon dioxide ice.[10] His observations that showed evidence of water ice were correct, though it took the planetary science community another half century and multiple space missions to Mars to correctly solve the riddle of the Martian polar caps.

Nevertheless, having solved the atmospheric problem correctly and the polar cap problem partially correctly, Kuiper was ready to move on to the next problem in need of a solution, "the nature of the green areas on Mars, often held to be vegetation because of the observed seasonal changes." Kuiper designed an experiment using observations at wavelengths of 0.6, 0.8, 1.0, and 1.6 microns, in which he observed both the green areas of Mars and the surrounding, so-called desert regions.

Human eyes are sensitive to light in the visible spectrum, from violet at the short wavelength end (at about 0.4 microns) to red at the long wavelength end (at about 0.67 microns). And chlorophyll, Kuiper knew, reflects light very effectively in the green (wavelengths close to 0.51 microns) and yellow (wavelengths close to 0.57 microns), i.e., at wavelengths shortward of 0.6 microns, and absorbs light at other visible-light colors. Green plants also reflect light very strongly at infrared wavelengths in the vicinity of 0.8 to 1.0 microns, a region in which human eyes are not sensitive. Kuiper hypothesized that if he could measure the color contrast between the green regions of Mars and the Martian desert regions he might resolve the question of what was producing the green color on Mars. If Earthlike, chlorophyll-containing plants were present on Mars, the percentage of reflected light from Mars would be dramatically lower at 0.8 and 1.0 microns than at 0.6 and 1.6 microns. But if no chlorophyll were present, the percentage of reflected light would be indistinguishable at the four different wave bands.

Kuiper found that the contrast between the two regions did not change across the four wavelength bands, a result that ruled out chlorophyll as the source of the green color on Mars. With this very simple and elegant experiment, Kuiper had ruled out Earthlike "seed plants" as the dominant kind of vegetation that existed on Mars. As explained again in *Time* magazine, "They could not be vegetation like trees or grass." He concluded that this was "not surprising, in view of the extreme rigors of the Martian climate, in particular the cold nights ... Seed plants and ferns are both vascular plants containing a great deal of water. Such plants would undoubtedly freeze in the Martian climate."

If not seed plants, what form of vegetative life could exist on Mars? *Time* explained Kuiper's ideas to readers this way: "They might be lowly lichens like those that grow on the dry rocks near McDonald Observatory. Lichens need no water in liquid form. Martian lichen-like plants might get enough water out of vapor from the ice caps, which evaporate without melting." Kuiper himself wrote, in his chapter "Planetary Atmospheres and Their Origin," in his own 1948 book, *The Atmospheres of the Earth and Planets*, "The hardiest terrestrial plant life are the lichens, a symbiosis of fungi and algae." Also, he made clear to his readers, the reflectance spectra of lichens are just like what he had found for Mars; that is, lichens show the same color contrast at 0.6, 0.8, 1.0, and 1.6 microns as does Mars. "The spectrum of these lichens and that of the Martian regions are similar between 0.5 and 1.7 μ [microns]."[11] While not saying, directly, that he had found spectra of lichens on Mars, Kuiper certainly meant to imply something very close to that idea. In fact, he presented a calculation that demonstrates that "if all atmospheric water vapor were made available to the green areas [which cover, he estimated, one-third of Mars], they would receive a layer 0.02 mm thick. The height of the living parts of the 'vegetation' could hardly be more than ten times this amount, or 0.2 mm; presumably it is much less. *This estimate is compatible with a lichen cover.*"* Kuiper continued his arguments in favor of lichens by noting that the green areas on Mars should long ago have been covered with yellow dust from the Martian dust storms unless they had the power to regenerate their green colors. He also noted that the complete absence of oxygen cannot kill terrestrial lichens, so we can feel confident that lichens could survive on Mars without oxygen. In addition, lichens "produce very little oxygen, and even the traces set free would gradually escape the planet." As a result, the fact that observers had not detected oxygen in the Martian atmosphere is consistent with lichen life on Mars. "No contradiction with the spectroscopic tests would thus result," he writes. Ultimately, Kuiper summed up the

---

*Italics in original text.

arguments in favor of lichens on Mars favorably, but with a level of caution not evident in most of his arguments: "Final judgment should probably still be withheld." Such caution, at least, was wise. If only he had taken his own advice.

Kuiper was careful in his later scientific papers in the *Astrophysical Journal* to merely imply that the spectra were consistent with the spectra of lichens and not to say that he had discovered lichens on Mars. He was perhaps a bit less cautious, however, when speaking with the reporter from *Life*, and later he found himself walking back his lichen claims just a bit.

In 1955, in the *Publications of the Astronomical Society of the Pacific*, Kuiper wrote that he did not think Mars had canals and that perhaps he had been misquoted about the whole lichen business, but not about the presence of some kind of plant life on Mars. "I did study Mars with the 82-inch telescope in 1948, 1950, and 1954, often under excellent conditions, with powers 660 and 900 times; and I have never seen a long, narrow canal nor a network of 'fuzzy canals.' I am personally convinced that the objective evidence which has led to this concept has been misinterpreted and erroneously represented on the drawings."[12] Having dismissed the existence of the canals, he then examined the evidence for the existence of vegetation. His conclusion, in 1955, was that though the canals did not exist, the evidence for vegetation probably did. That evidence was ambiguous but likely; in fact, his discoveries provided the weight of the evidence in favor of the biological hypothesis. In his own words, "Seasonal and secular variations in the extent of the dark areas have been described by several authors. . . . These variations have often been regarded as strong indication that the dark areas are covered with vegetation; but they are, of course, insufficient as proof. An inorganic explanation cannot be immediately ruled out, although such explanations as have been advanced appear improbable. The discovery of CO_2 in the Martian atmosphere at the McDonald Observatory in 1947 and the infrared spectrum of the polar cap showing it to be frozen H_2O not CO_2, have greatly enhanced the a priori probability that some primitive vegetation exists on Mars."

Given that he had embraced "primitive" vegetative life on Mars as a much more likely explanation for the color changes of the dark areas than weather effects or areology,* he addressed his earlier statements about lichens. "The hypothesis of plant life, for reasons developed elsewhere, appears still the most satisfactory explanation of the various shades of dark markings and their complex seasonal and secular changes. I should, however, correct the impression that I have supposed this hypothetical vegetation to be lichens. Actually, I said: 'Particularly, the comparison with lichens must be regarded to have only heuristic value; it would be most surprising if similar species had developed on Mars as on the Earth.'" Indeed, these words did appear exactly as he quoted himself, near the end of his "Planetary Atmospheres and Their Origin" chapter in his 1948 book *The Atmospheres of the Earth and Planets*.

By the mid-1950s, with multiple observers on several continents working across three decades and none of them finding any evidence for chlorophyll on Mars, and with Lowell's canals all but forgotten, Kuiper's "primitive vegetation" was now the only remaining scientific leg on which the life-on-Mars argument still stood. Kuiper himself was becoming very careful in his professional writings, reminding his colleagues that his lichens-on-Mars idea was merely a hypothesis advanced to explain the seasonal and long-term changes in the dark markings on Mars, and he refrained further from any bolder statements to reporters from *Time* or *Life* magazines.

Writing in the March 1957 issue of the *Astrophysical Journal*, he backtracked a bit more when he noted that "The lack of vivid colors observed during the 1954 and 1956 oppositions (early and late spring, respectively, in the southern Martian hemisphere) suggests that an inorganic explanation for the dark markings be considered along with the vegetation hypothesis. . . . The most probable inorganic hypothesis would appear to be that the maria are *lava fields*, somewhat like the lunar maria and perhaps those of Mercury. On general grounds, such a unifying hypothesis would be attractive. . . . Apparently, while the sand fills the crevasses in the lava, it blows off the vitreous surface. Lava fields therefore have the 'regenerative

---

*Geology on Mars.

power' after a dust storm that has been invoked as an argument favoring the vegetation hypothesis."

Kuiper, though becoming more cautious, was unable to completely give up on the idea of Martian plant life. He stated in his very last sentence, "As a working hypothesis it is supposed that the maria are lava fields that may have a partial cover of some very hardy vegetation."[13] Kuiper, having backed away from asserting that he had discovered lichens on Mars, had covered all his bases.

The momentum had shifted away from the belief that our neighboring planet Mars, the pale red dot in the nighttime sky, was a planet brimming with life. Then, just when the idea of Martian life might have been extinguished for good, Bill Sinton showed up.

algae

One of the unfortunate legacies of the flawed work of Percival Lowell was that planetary astronomy was not a popular pursuit in the first half of the twentieth century. Instead, most astronomers studied stars and galaxies (and had great success in learning about the physics of stars and the universe). One had to be a bit foolish and quite a bit courageous to decide to pursue a career in this neglected backwater. Bill Sinton had fought with the Twenty-sixth Infantry Division in the Second World War. That difficult experience may have served him well when he decided to follow Kuiper's lead and become one of the pioneers in developing the field of infrared astronomy and applying that approach to studying objects in the solar system. For his doctoral dissertation research at Johns Hopkins University, he measured the infrared spectrum and the temperature of Venus and other planets. He next directed his attention to the Moon and Mars.

One of the biggest obstacles he had to overcome was the enormous amount of infrared light (the "background signal") that was emitted by the detector he was using to measure the infrared light from distant, astronomical objects. All objects emit light, and the kind of light they emit most effectively depends on their temperatures (this is known as "blackbody radiation"). Objects with temperatures of millions of degrees emit light most effectively in the form of X-rays. Objects like the Sun, with temperatures of thousands of degrees, emit visible light best. Objects at room temperature (hundreds of degrees Kelvin), like buildings, telescopes, astronomers, and astronomical detectors, are prolific emitters in the infrared. At temperatures below about 100 degrees (Kelvin), objects emit microwaves or radio waves prolifically but give off almost no infrared, visible, ultraviolet, or X-ray photons.

Sinton knew that he could decrease the intensity of the problematic infrared background signal by reducing the temperature of the detector. In principle, when the detector was cool enough the infrared light emitted by the detector would approach a negligible level; consequently, detecting the infrared signal from the Moon or Mars would become possible. Working with the 61-inch telescope at the Harvard College Observatory (installed in 1934), he used an instrument he built by himself, one that was cooled by liquid nitrogen down to almost 300 degrees below zero (−287°F, or 96 K), a remarkably cold temperature for an astronomical detector system at that time. At such a low temperature, infrared studies of Mars became possible.

Sinton decided to study Mars at the infrared wavelength of about 3.4 microns in order to search for signs of vegetation. Thanks to Kuiper, as was well known, he wrote, "there is already important evidence pointing toward the presence of vegetation on Mars."[14] Using his pioneering skills in infrared astronomy, he was going to gather substantive, spectroscopic proof.

At wavelengths from 3 to 4 microns, Mars emits much less light of its own than the amount of infrared sunlight it reflects. Material on the Martian surface, however, could have an effect on the spectrum of the reflected sunlight. In particular, laboratory work published in 1948 by a team of chemists had shown that when two carbon atoms each share an electron with a single hydrogen atom, as would be expected in organic* molecules, that combination of atoms is especially good at both absorbing and emitting light at a wavelength of 3.46 microns. If the organic molecule is very lightweight, for

---

*An organic molecule must contain one or more carbon atoms and must have carbon-hydrogen (C-H) bonds.

example a methane molecule, which is made up of only one carbon atom and four hydrogen atoms (CH_4), the band at which the
molecule is an effective absorber and emitter of light can shift to
a wavelength as short as 3.3 microns. In general, larger and heavier
organic molecules, like those in biological materials on Earth, shift
this spectral feature into the 3.4–3.5-micron range. Sinton designed
an experiment to study Mars and look for this spectral feature, under
the assumption that plant life on Mars would show this same signature as plant life on Earth.

If we imagine that an incoming source of light (e.g., sunlight) includes approximately the same amount of light at 3.1, 3.2, 3.3, 3.4,
3.5, and 3.6 microns, then when light waves at all of these distinct
wavelengths reflect off a surface that includes leaves or moss or grass
or lichens, the organic material, because it is full of those C-H bonds,
absorbs light effectively in the 3.4–3.5-micron range and reflects light
effectively at the other wavelengths. If we were to plot the intensity
of reflected light (the y-axis value) as a function of the wavelength of
light (the x-axis value), we would see a constant amount of light at
3.1, 3.2, 3.3, and 3.4 microns, a drop in the light intensity from 3.4 to
3.5 microns, and a rise in the light intensity from 3.5 to 3.6 microns.
This drop in the light intensity across a short range of wavelengths
is an absorption band, similar to those Fraunhofer bands first discovered a century and a half earlier. Sinton obtained infrared spectra of
terrestrial biological materials in order to demonstrate the presence
of this absorption band in light reflected from living things, including a lily of the valley leaf, a maple leaf, two types of lichen, and a
moss, all of which did show some sort of absorption feature in the
3.4–3.5-micron range. Then he collected infrared spectra of Mars and
compared them to his test spectra.

The result? The Martian spectra he obtained in late 1956 showed
"a depression at the wave length of the organic band" at 3.46 microns. In a paper he published in the *Astrophysical Journal* in 1957,
he wrote that this dip did not prove that lichens are present but did
indicate "that organic molecules are present."[15]

Sinton was correct in asserting that the presence of a bona fide absorption band at this wavelength would demonstrate that some material that is good at absorbing light at 3.46 microns exists somewhere

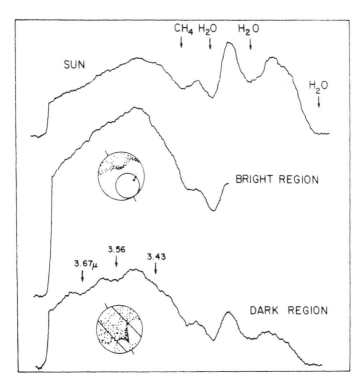

Figure 9.2. A plot that compares the amount of reflected light from the Sun (top), a bright region on Mars (middle), and a dark region on Mars (bottom), as a function of wavelength. Long wavelength (about 4 microns) is to the left; short wavelength (about 3 microns) is to the right. The apparent drops in the intensity of reflected light at 3.67, 3.56, and 3.43 microns in the dark regions of Mars, in contrast to the absence of these apparent drops in both the bright regions of Mars and in direct sunlight, were interpreted by Sinton as evidence for vegetation on Mars. Image from Sinton, *Lowell Observatory Bulletin*, 1959.

along the light path between the Harvard College telescope in Massachusetts and the surface of Mars. He was on weaker ground in arguing that that material was organic. Whatever was absorbing the light in the 3.46-micron band might be in Mars's atmosphere, but it also could be in Earth's atmosphere. Sinton quickly concluded, however, that he had more knowledge about that absorbing material than he, in fact, actually had. He claimed that "the dip at the significant wave length is therefore additional evidence for vegetation. This evidence, together with the strong evidence given by the seasonal changes, makes it extremely likely that plant life exists on

Mars." Unlike Kuiper, Sinton did not take a cautious approach in his professional writing. He jumped in with both feet, concluding without equivocation that he had uncovered clear evidence for Martian vegetation.

Two years later, when Mars returned to a position in the sky where it was again well positioned for viewing from Earth, Sinton returned to his studies of Mars. In 1958, the steady advance of technology offered Sinton the benefit of improved equipment. In addition, his first success in supposedly detecting life on Mars was leverage for moving his observing program from the 61-inch Harvard Observatory telescope to the biggest telescope on the planet at that time, the 200-inch Hale telescope on Palomar Mountain in southern California, which had opened in 1948. For this project, he was awarded two weeks of use of this telescope during the time of the month astronomers call "bright time," this being the nights just before, during, and after the night of the month when the Moon is full and when moonlight illuminates much of the night sky.

This approach is quite typical in astronomy. Use your initial discovery on your "small" local telescope as leverage to obtain access to observing time on a much bigger telescope. With a bigger telescope, one collects more light from the astronomical target in the same amount of observing time than one would with the smaller telescope, thereby gaining efficiency and a more reliable scientific result. In addition, bigger, newer telescopes tend to be placed on mountains with better weather conditions and usually have more modern equipment than smaller, older telescopes. As a result, with the bigger telescope, in this case with the biggest telescope in the world, Bill Sinton expected to carry out the definitive observing program in the study of Mars and prove the existence of plant life on Mars.

Sinton's goal in 1958 was to remove any doubt about the reality of the absorption band he had claimed to have identified in his 1956 observations. In his own mind, without any reservations, he achieved that goal: "the reality and distribution of the band were established." His new discoveries were published in 1959 in the prestigious scientific journal, *Science*, which maximized the impact of his work and generated the most fame for himself and publicity for his work.[16]

According to Sinton, the Martian spectra, in particular those taken when observing the dark patches on Mars, revealed absorption bands that were strongest at 3.43 and 3.56 microns and showed a third absorption feature at 3.67 microns (later revised to 3.45, 3.58, and 3.69 microns[17]). The "bands near 3.5 microns" are "most probably produced by organic molecules," he wrote, and these are "produced in localized regions in relatively short spans of time. Growth of vegetation certainly seems to be the most logical explanation for the appearance of organic molecules."

The longest wavelength band had Sinton puzzled, because "it had not been seen in any terrestrial plants." However, perhaps through serendipity, he conducted laboratory spectral tests of the lichen *physica* and the alga *cladophora*. Both *physica* and *cladophora* showed a shallow absorption feature at about 3.7 microns, in addition to deeper absorption bands at about 3.43 and 3.56 microns.

Sinton concluded that the similarity of the Martian spectrum to that of *cladophora* demonstrated that what was observed on Mars was "produced by carbohydrate molecules present in the plant. The attachment of an oxygen atom to one of the carbon atoms shifts the resonance of a hydrogen atom attached to the same carbon to a longer wavelength. Thus, the evidence points not only to organic molecules but to carbohydrates as well." Sinton suggested that this spectral signature was not only proof for the presence of life on Mars but was also a clue about the need for Martian plants to have large food storage capacities.

Sinton had hit an apparent grand slam. He had not just uncovered evidence for life on Mars. He had discovered spectral features that revealed what kind of life exists on Mars. Not Kuiper's lichens. Instead, Martian algae very closely resemble terrestrial algae that produce and store carbohydrates. Soon, other scientists would refer to the 3.4–3.7-micron spectral features on Mars as the Sinton bands, and *Science* magazine would become the town square for arguments for and against the Sinton bands as proof of evidence for life on Mars.

Notably, in 1958 Sinton had obtained solar spectra in the afternoon by observing light reflected off a piece of aluminum (which was thought to be a near-perfect reflector). Presumably, Sinton assumed he could use the reflected-light spectrum of aluminum to correct the Martian spectra for any spectral effects due to the Sun or to Earth's atmosphere, but these spectra perhaps were insufficient for this task. He did not obtain (nighttime) lunar spectra, with the Moon at a similar height in the sky as Mars, as a comparison, perhaps because, based on his 1956 observations, he had concluded that a (reflected) solar spectrum was a better calibration source than a lunar spectrum. In actuality, a lunar spectrum probably would have done a better job in correcting for terrestrial atmospheric effects, and if he had used this method, he might have discovered his error sooner rather than later.

In a letter to *Science* published in 1961, Norman Colthup, of Stamford Research Laboratories, in Connecticut, and a giant in the field of infrared spectroscopy, wrote to agree that the band seen by Sinton at 3.43 microns was almost certainly due to a carbon-hydrogen bond in "carbohydrates and protein organic matter in plants which resembled terrestrial plants."[18] Colthup goes on to say that the only likely source of the two spectral features at 3.56 and 3.67 microns must be molecules known as "organic aldehydes (but not formaldehyde)," because organic aldehydes "are among the few materials with a strong band near 3.67 microns" in addition to the band at 3.43 microns.

Aldehydes are chemical compounds that contain the elemental group CHO, in which a carbon atom shares two electrons with an oxygen atom and shares a third electron with a hydrogen atom. Because a carbon atom has four electrons available for sharing, the carbon atom in the CHO group still has one more electron bond it can share with yet another atom. How that fourth electron is shared determines the actual chemical species for this compound (e.g., is it methanal, ethanal, or propanal?). Aldehyde is in the ethanal family. As Colthup noted, acetaldehyde is a very effective absorber at wavelengths very close to 3.58 and 3.68 microns, which, he argued, makes it a great match to the Martian spectral signatures.

Colthup identified the specific aldehyde, acetaldehyde (or ethanal, C_2H_4O), as the most likely source of the Sinton bands and attributes the presence of this material on Mars to the near absence of oxygen on Mars, since this particular molecule preferentially forms in an oxygen-poor environment. Just in case he had not gone too far already, Colthup continued by going much too far: "If I may be permitted to speculate a bit, acetaldehyde may be an end product of certain anaerobic metabolic processes." He pointed out that one such process is fermentation, in which carbohydrates are converted to acetaldehyde and finally to alcohol. "This process yields much less energy for the organism than conventional oxidation … but certain organisms on Earth use fermentation as their source of energy when oxygen is not available, and perhaps this happens on Mars."[19]

As often happens in the competitive, scientific marketplace of ideas, not everyone agreed with Colthup and Sinton. In 1962, Donald G. Rea, of the Space Sciences Laboratory at the University of California at Berkeley, began the attack. In a major review of the studies of planetary atmospheres carried out by astronomers up until that time, he pointed out that his own calculations showed that at the temperatures and pressures of the Martian atmosphere at the Martian surface, the vapor pressure should force acetaldehyde molecules into the gas phase and thereby into the atmosphere.[20] In fact, "the high volatility of this chemical would ensure a high concentration in the atmosphere … It should then be observed over the entire disk [of Mars] and not be restricted to certain areas." Sinton, however, saw these supposed absorption bands in reflected light only from the dark areas on Mars. Since the absorption bands were not evident from the light areas of Mars, Rea argued that the conclusion drawn by Colthup was wrong. Rea knew that Sinton's flawed conclusions were vulnerable to his laboratory work and, like a dog with a bone, he was not about to let go.

A year later Rea and two of his colleagues in the Department of Chemistry at Berkeley, T. Belsky and Melvin Calvin, presented a large body of laboratory spectra that revealed additional problems with Sinton's work.[21] Some of these problems were instrumental. For example, when light from Mars was focused by a telescope onto Sinton's detector, the light had to first pass through a window made of Plexiglas in order to enter the sealed, supercooled device that holds the infrared detector. Rea and his colleagues asserted that the measured wavelength of the light wave changed as a result of being transmitted through the Plexiglas. The result of using the Plexiglas window, they claimed, is that the experimenters might not be able to know the correct wavelength of the spectral feature on Mars before it was transformed.

Other problems had to do with which absorption bands are properly identified with the alga *cladophora* or with other biological materials Sinton tested, including lichens and the lily *agapanthus*. The Berkeley chemists showed that Sinton's assignment of a single spectral feature to *cladophora* was a marginal conclusion, at best, and that the Mars spectra were missing other spectral signatures of *cladophora*. If they have one of the *cladophora* spectral features, shouldn't they also have the others, they asked? They knew the answer was "yes." Perhaps, then, the evidence for *cladophora* was not as definitive as Sinton claimed. In 1963, Rea, Belsky, and Calvin mostly raised questions but provided no firm answers. The questions, however, were a sign that trouble was brewing for Sinton and his supporters.

Another team of Berkeley chemists, James Shirk, William Haseltine, and George Pimentel, were similarly unconvinced that the Sinton bands were evidence of the presence of algae or of active fermentation processes on Mars. They went looking for other possible explanations. In January 1965, they found one: water. In fact, the explanation they found centered on the presence of both heavy water and semi-heavy water in the Martian atmosphere.

Just as the three isotopologues of water (H_2O, HDO, and D_2O) respond differently to the force of gravity, they also absorb and emit light at slightly different wavelengths. Shirk and his colleagues found that HDO and D_2O absorb light at wavelengths that match the wavelengths of the Sinton bands more closely than does acetaldehyde. In other words, "the reported absorptions could be attributed to HDO or D_2O in the Martian atmosphere."[22] Their explanation raised other interesting questions in need of answers. Why, for example did the dark areas on Mars show a higher water vapor content than the light areas?

In March 1965, the algae and the fermenters on Mars finally met their Waterloo in Denver, Colorado. Using observations of the Sun obtained at the University of Denver and measurements of the total amount of water vapor that would have been in the air above Palomar Mountain in Earth's atmosphere when Sinton used the Hale Telescope to make his original observations of Mars in the late 1950s, the team of Rea and Brian T. O'Leary, together with Sinton himself drew the conclusion that "there seems to be a correlation between the intensities of the 3.58 and 3.69-micron features and the amount of telluric water vapor in the optical path. An important corollary is that there is no evidence for attributing these spectral features to Mars."[23] That rather quiet, low-key sentence bears repeating: *there is no evidence for attributing these spectral features to Mars*. The Sinton bands at 3.58 and 3.69 microns are telluric! They have everything to do with the water content of Earth's atmosphere and nothing to do with anything in Mars's atmosphere. Sinton had detected water, specifically heavy water, on Earth, not life on Mars.

What about the third Sinton band at 3.45 microns? In hindsight, an impartial referee who might examine the original spectrum published by Sinton in his 1959 paper and who looks for the unexplained spectral feature at 3.45 microns is unlikely to find it. No such feature ever existed. No such absorption band ever needed to be understood. All Sinton had measured was noise.

To his great credit, Sinton was one of the co-authors of the paper that demolished a decade of his own work. Such a high-profile, public acknowledgment of such a major scientific error is rare and brave. Thereafter, Sinton had a long and moderately distinguished scientific career. He helped to establish the observatories on the summit of Mauna Kea in Hawai'i for the University of Hawai'i, studied the volcanoes on Jupiter's moon Io, and conducted extensive infrared studies of the Moon and the atmospheres of the planets Uranus and Neptune.

Science magazine had been the purveyor of the yin and yang of the debate about the Sinton bands for a decade. Now that debate was over and a new era in planetary science had begun when NASA's Mariner 4 spacecraft, launched in November 1964, returned the first-ever close-up image of Mars on July 15, 1965. During the Mariner 4 flyby, the spacecraft took twenty-one pictures of Mars. To commemorate the one-year anniversary of that historic event, in the July 15, 1966, issue of *Science*, the magazine turned to Ernst Julius Öpik to offer some perspective on Mars. Estonian by birth, Russian by education, Öpik finished his career at Armagh Observatory in Northern Ireland, where he edited the *Irish Astronomical Journal*. He was inducted into the United States' National Academy of Sciences in 1960 and received the Gold Medal from the Royal Astronomical Society in 1975. Öpik was one of the giants of astrophysics and planetary astronomy in the twentieth century, and he had significant weight to toss around when he explained what astronomers knew about life on Mars for *Science*.

Öpik summed up what we Earthlings thought we knew at the time about Mars in his twelve-page article, "The Martian Surface."[24] "The story of the infrared bands near 3.6 microns is very instructive," he wrote, pointing out that the scarcity of our observations is problematic. "These bands were at first attributed to the CH bond characteristic of complex hydrocarbons and indicative of some kind of organic matter. Then it was shown that they fitted absorption bands of heavy water, HDO, much better. This led to speculations on the enrichment of deuterium on Mars, through preferential escape to space of the lighter hydrogen, or through some other mechanisms. Finally, it turned out that the bands belong to heavy water in the terrestrial atmosphere and have no bearing on Mars." Clear, succinct, and accurate.

The surface of Mars, he continued, has regions of different colors. About 70 percent (the continents) of the surface can be characterized as orange, yellow, or red. The rest of the regions (the maria) are darker, and are "sometimes described as greenish or bluish, but they are only less reddish" than the other areas. Here, finally, Öpik had pulled back the curtain and acknowledged that Mars had never looked green. It only sometimes looks a bit less red. And, as has been known for centuries, the shades and colorations change with the seasons.

How are we to understand all of this, he asked? Despite all the evidence to the contrary, despite the barrenness of Mars revealed by the admittedly grainy Mariner 4 images, despite the implosion of the Sinton bands as evidence of Martian life, despite his acknowledgment that Mars has never shown any green colors, but is only sometimes less red, Öpik, like so many astronomers before, found himself trapped by the weight of astronomers' long love affair with the idea of Martian life. Öpik's answer: "Against this background, the survival and permanence of the Martian surface markings seem to suggest a certain regenerative property. . . . Vegetation growing in favorable places on the drifting dust could serve as an explanation. . . . That vegetation could survive the extreme cold and dryness of the Martian climate may seem incredible. Yet the very rigor of the environment may help in this respect: the nocturnal temperatures are so low that hoarfrost may be deposited (as is apparently observed on the soil and hard-frozen plants; when this melts, drops of liquid water may be utilized by the plants while they are warming up in the morning sun. . . . However doubtful the vegetation hypothesis of the maria may appear, it is difficult to find an alternative that accounts for all the facts."

No evidence for chlorophyll. No proof of lichens or primitive vegetation. No CH bands due to algae. No "green" areas. But still the idea of life persisted. By 1965, Mars was also known to be incredibly cold, oxygen-deprived, and nearly water-free. Because the Martian atmosphere is so thin and has no ozone layer, the surface is exposed to lethal levels of ultraviolet light from the Sun and is bombarded by deadly cosmic rays from the Sun and beyond. Yet, despite all of those clear reasons to dismiss vegetation as a possibility on the Martian surface, Öpik couldn't do it. The desire to find life on Mars was too deeply ingrained in our mid-twentieth-century human psyche for Öpik to let go of that possibility.

Two giants of late twentieth-century planetary astronomy, James Pollack and Carl Sagan, finally put this issue to rest with their work a year later, in 1967. Sagan was already well known for studies of the atmospheres of Mars and Venus and for his speculative book *Intelligent Life in the Universe*, published in 1966, though not yet world-famous for his work on the Viking missions and his narration of the *Cosmos* television series on National Public Television. Pollack would go on to play important roles in virtually every NASA mission to every planet for decades, including Mariner 9 (to Mars), Viking 1 and 2 (both to Mars), Voyager 1 and 2 (to Jupiter, Saturn, Uranus, and Neptune), Pioneer Venus, Mars Observer, and Galileo (to Jupiter). He would become an expert on the physics of Saturn's rings, on the formation of giant planets, and on the evolution of Earth's early atmosphere.

Pollack and Sagan used radar maps of Mars, collected with one of NASA's deep space network antennas, the Goldstone tracking station in California, managed by the Jet Propulsion Laboratory, to study and understand the surface of Mars. With one of the Goldstone dishes, they sent out radio waves that bounced off the surface of Mars; they used the same giant dishes to measure the reflected signals. The dark areas of Mars, they discovered, were at higher elevations than the bright areas. In addition, the dark regions that exhibit long-term changes in coloration have smaller slopes and elevations than the dark regions that do not undergo secular changes in color. As a result, they "hypothesize that the secular changes are due to the movement of sand and dust from the bright areas onto and off from adjacent dark areas of shallow slopes." The so-called regenerative properties of the dark areas, that is, the ability of the dark areas to darken after lightening, "are due to winds scouring small deposited particles off the sloping highlands."[25]

In other words, Martian weather systems drive windblown sand across the surface. The sand grains are preferentially deposited in the lowland "deserts." The sand grains are highly reflective, so the deserts look fairly bright. When the small sand particles are blown onto and cover the shallow, sloped surfaces, these regions brighten and look like the deserts. When the winds scour the small particles of sand off the shallow slopes, these sloped surfaces rapidly darken. The darkening of these surfaces had been interpreted (and misunderstood) as the rapid regeneration of plant life in Martian springtime.

Sagan and Pollack later presented a quantitative model for windblown dust particles on Mars to support their initially, purely descriptive hypothesis.[26] While they concluded that "the success of windblown dust models does not, of course, argue against life on Mars," the success of their windblown dust models did finally put an end to the idea that the changing colors on Mars are caused by waves of greenery that roll across Mars as melting water from the polar caps ushers Martian springtime into bloom. No more trees; no more moss; no more lichens; no more algae. Just windblown sand.

10

---

vikings on the plains of chryse and utopia

On July 20, 1976, Viking 1 landed on the Plains of Gold, Chryse Planitia, on Mars. Six weeks later, on September 3, Viking 2 touched down
on Utopia Planitia.* Not only had both of the missions, launched
from Cape Canaveral in August and September 1975, successfully
reached Mars, both of the landers touched down safely on the surface. With the same suite of scientific instruments on board, the Viking 2 Lander touched down at a location about 25° farther north
on Mars and halfway around the planet in longitude from Chryse
Planitia. Now, in two different locations, human-directed science experiments would take place for the first time on the surface of Mars.

Before Viking 1 and Viking 2 landed on Mars, humanity's collective knowledge of Mars was extremely limited. Though no longer trapped by Percival Lowell's vision of a Mars with intelligent
beings who had built canals to transport water from the water-rich
poles to the arid equator, Mars remained mysterious. Planetary scientists knew the Martian atmosphere was thin and mostly made of
carbon dioxide. Astronomers had learned that first from the telescopic studies of Gerard Kuiper and had enhanced and improved on

---

*The names Chryse Planitia and Utopia Planitia were both approved by the IAU in 1973; they
derived from classical usage, as found in Antoniadi's early-twentieth-century maps of Mars.

that knowledge with the detailed data from the Mariner missions. Also from the Mariner imaging surveys, the surface, we then knew, was covered with cratered plains, a small number of enormous volcanoes, and a significant number of narrow features cut into the surface that looked like ancient, dried-out riverbeds and lengthy outwash channels. At that time, the polar caps were thought to be composed mostly of frozen carbon dioxide, though planetary scientists also knew that in Martian summer, after the CO_2 sublimated into the atmosphere, a residual polar cap of water ice remained.

Mostly, however, scientists had more questions than answers about Mars. Was the surface covered with a deep layer of soft dust? Would the landers fall over after touchdown? Would they sink into the powder? Most geologists were expecting to find a fairly hard surface and would have bet against these more pessimistic scenarios for a landing fiasco, but mission planners wouldn't know the answer until the first Viking lander settled onto the surface. As they found out, the dust on the Martian surface was not that thick and powdery; to the relief of everyone at NASA, the landers did not fall over.

Did the atmosphere contain any nitrogen, let alone enough nitrogen to sustain life? Nitrogen, after all, is critical to life on Earth. It is a major elemental component of both amino acids and the nucleotide bases that are fundamental to the makeup of DNA. Although astronomers had been certain in the 1950s that the Martian atmosphere was more than 95 percent nitrogen, those early assumptions had been dashed by actual measurements, and, because Mariner 9 had been unable to detect nitrogen in the atmosphere of Mars, pre-Viking planetary scientists knew the nitrogen abundance in the Martian atmosphere was at most a few percent. In one of the most important astrobiological discoveries ever made about Mars, the Viking mission confirmed that nitrogen does exist in the lower Martian atmosphere, at a level of about 2.7 percent.[1] (In 2013, a measurement by the Curiosity rover team[2] incorrectly placed the nitrogen abundance at 1.89 percent. This result has since been "reassessed." The revised nitrogen abundance measurement from the Curiosity team,[3] 2.79 percent, very closely matches the Viking lander measurements.)

Mars was known to have mighty, planet-girdling dust storms. Might these storms generate hurricane-force winds that could blow over the landers after a successful landing (as they did in the movie *The Martian*)? While the storms lift enough dust into the atmosphere to block our view of the surface for months at a time, the answer is no: the atmosphere is so thin that it lacks the momentum to knock over the landers.

Might "macrobes" exist? The word macrobes was Carl Sagan's imaginative term for life-forms visible to the naked eye. If they existed on Mars, might they appear in images obtained by Viking at the landing sites? Sagan was not crazy, just open to all possibilities that could not be absolutely denied. "In fact," he said, "there is no reason to exclude from Mars organisms ranging in size from ants to polar bears. And there are even reasons why large organisms might do somewhat better than small organisms on Mars."[4] Though Sagan seemed to be channeling Lowell, he most likely didn't think the likelihood that macrobes existed on Mars was high, but before the Viking landers arrived on Mars, nobody could say, with absolute certainty, that macrobes were absent on Mars. The experiment of looking for macrobes had to be done, Sagan insisted. Cameras had to land on Mars and examine the environment for macroscopic Martians. He won that argument, and so both Viking landers included cameras capable of scanning the panorama to look for Martians. These same cameras were also available for use in fundamental geological studies.

After Viking 1 and 2 each landed, both spacecraft almost immediately returned images of the Martian surface to eager eyes on Earth. No macrobes peered back at Sagan through the eyes of Viking. The search for life on Mars could now concentrate on the Viking biology experiments designed to search for evidence of microscopic life. These experiments commenced on the eighth Martian day after landing, designated Sol 8. Three days later, the weirdness began.

On July 31, 1976, Sol 11, Harold Klein, the director of Life Sciences at NASA's Ames Research Center and leader of NASA's biology investigation team for the Viking mission program, spoke to the world. The underlying premise, the foundation, for this press conference had been set years before when NASA selected these investigators and the investigations they led for the Viking missions and packaged them together under the label of "biology experiments." Klein and the other biology investigation team members were looking for

Figure 10.1. First Viking lander survey panorama, obtained on the surface of Mars on July 20, 1976, revealing no macroscopic Martians. Image courtesy of NASA.

evidence of life on Mars, and this press conference was the first opportunity for the members of the biology investigation team to report on their results. Had they found life on Mars?

Klein spoke deliberately, but optimistically. One of the two Viking biology experiments, the Gas-Exchange experiment, had already yielded "at least preliminary evidence for a very active surface material." Those words were intentionally selected to suggest to listeners and reporters that NASA had probably found life on Mars but were not about to say so directly. A second experiment, the Label-Release experiment, had generated a response that looked "very much like a biological signal." Those were even stronger and less cautious words. But Klein quickly backpedaled. Both results, he cautioned, "must be viewed very carefully." Klein continued, trying to stay on message, "We believe there is something in the surface, some chemical or physical entity, which is affording the surface material a great deal of activity and may in fact mimic—let me emphasize that: mimic—in some respects, biological activity."[5]

Klein began this discussion with the press and the world about the biology experiments under way on the surface of Mars, through which he and his colleagues were searching for signs of life on Mars, with words that he felt were ones of great circumspection, caution, and reasonableness. He made extremely clear the preliminary nature—in hindsight, we could say ridiculously premature nature—of the biology team's early public conversation about their findings. But then Klein, like most scientists in these exchanges with reporters for which they are not trained and which are unlike scientific arguments with professional colleagues, almost immediately lost control of the conversation and the message. In reality, the conversation was never his to control. The *New York Times* headline on August 1 screamed from page 1, "Scientists Say Data Could Be First Hint of Life on the Planet."[6] Going forward, the press would question whether the Viking experiments could possibly *dis*prove the existence of life on Mars.

One reporter asked whether the production of oxygen in one of the experiments was evidence for photosynthetic activity on Mars. The reporter who asked this question either had not invested the time and energy, beforehand, in trying to understand the details of the experiment about which he was asking a question, or did not understand how photosynthesis works. Perhaps this reporter slept through science classes in junior high school. Yes, during photosynthesis plants breathe in carbon dioxide and exhale oxygen, but they do so using energy from the Sun to power the process. On Mars, the Gas-Exchange experiment that had yielded some oxygen was taking place in absolute darkness inside the Viking 1 lander. Therefore, the simple answer to the question was "no." This experiment could not test for any kind of reaction similar to photosynthesis.

Another reporter asked Klein whether animals could have produced the oxygen. This reporter apparently skipped biology class. Animals, at least all of the animals on Earth with which both the reporter and the mission scientists were familiar, inhale oxygen and exhale carbon dioxide.

The press corps was not distinguishing itself as scientifically well-informed. The science team, or NASA, could have, and perhaps should have, educated the press corps better before inundating them with raw, barely evaluated data. Part of the mission protocol, however, was to immediately put new and exciting results—describing these results as "scientific" would be a stretch since they had barely, if at all, been vetted through any kind of scientific review process—in front of the eager public who had paid for this mission. The science team should have already sensed that the press conference wasn't going to go well and the science reported to the general public from the press conference likely wasn't going to be reported the way the scientists intended. Of course, if they had listened to their own arguments made in their own meetings earlier that same day, they never would have presented the message that included the words "very much like a biological signal."

"Tests by Viking Strengthen Hint of Life on Mars," blared another page 1 headline in the *New York Times* one week later, on August 8.[7] By now, the press controlled how the Viking mission science was interpreted and presented to the public. The Viking scientists would now have a hard time convincing the public that experimental evidence that strongly argued against the presence of life on Mars was anything more than ambiguous in meaning. An August 14 *New York Times* article reported that a soil sample result contained "no detectable level of complex carbon-containing molecules that might be produced by microbes" and that therefore that result was "inconclusive."[8] On August 21, the headline, relegated to page 18 in the *Times*, was "Experiment Fails to Rule out Possible Biological Processes on the Planet."[9]

The results generated by the Viking 2 lander experiments—an identical suite of instruments led by the same teams of scientists—did not receive the same attention from the public as did initial results of the Viking 1 lander experiments. Though Viking 2 landed only six weeks later, the biology investigation team leaders and the science team members were vastly wiser, as well as more experienced with their own experiments and with the press. The science teams mostly returned to their labs and their daily returns of data and tried to figure out how to best understand their scientific haul. Without the intense, inquisitive press corps prying into their work on a daily basis, the scientific teams carefully and methodically worked through their data, designing new experimental protocols in real time to test new hypotheses.

With the eventual passage of years, and under intense pressure from scientific colleagues, NASA, and the media to generate good scientific results, the scientists running the three separate Viking biology experiments tested and retested, calibrated and recalibrated, and argued and reargued. In the end, in contrast to the excitement about the possibility of discovering Martian life evident in that first press conference, almost all of them concluded that their combined results "indicated that extant organisms were not present at the two landing sites."[10] One team member, however, remains to this day convinced otherwise. He continues to argue that his experiment did, in fact, detect evidence of Martian biological activity.

In order to initiate the Gas-Exchange experiment, designed by a team led by Vance Oyama, of NASA's Ames Research Center, a robotic arm reached out from the lander and scraped a small amount of dirt off the surface of Mars, which it then dropped into a chamber inside the lander. (This same robotic arm would provide samples to the other Viking lander experiments that required Martian dirt.) Inside this chamber, the Martian dirt was in the dark, sealed off from the Sun, the external Martian environment, and all other pieces of Viking lander machinery.

The dirt in the sealed chamber was first tested without the addition of any water. Then it was exposed, indirectly, to water. The water was injected into a dish located below the soil, though the water and soil were never in direct contact. Moisture, evaporating off the water dish, penetrated and moistened the dirt from below. In the last phase of the Gas-Exchange experiment, a "chicken soup" mixture containing nineteen different kinds of amino acids was added to the dish of water. The chemically enriched moisture transported nutrients upward into the soil. If any living things were in the soil, they were expected to grow, respire, and reproduce. Presumably, they would exhale carbon dioxide, carbon monoxide, methane, hydrogen, nitrogen, oxygen, nitrous oxide, and hydrogen sulfide in detectable quantities. Amazingly, only 2.5 hours after the first injection of nutrients into Oyama's first dirt sample, the soil began to release large quantities of oxygen. The oxygen was released in a large burst, just as the experiment began. Then the oxygen level relaxed back to nominal levels. Was this burst of oxygen evidence of living creatures in the Martian soil?

Gilbert Levin's Label-Release experiment shared some similarities with Oyama's Gas-Exchange experiment. Levin, working for Biospherics, Inc., of Rockville, Maryland, had modified for use on Mars a device he had first invented for detecting microbes in polluted water. His device was designed to detect the carbon the microbes would emit as a result of their metabolic activities. On Mars, Levin's device, like Oyama's tiny bit of machinery, injected nutrients into Martian dirt, but Levin's nutrients were synthesized amino acids and carbohydrates, all of which contained atoms of radioactive carbon-14 in place of stable carbon-12 atoms. The detector in Levin's experiment was a Geiger counter that would detect the radioactive emissions from any carbon-14 atoms if they were eaten, or metabolized, by a living organism and then exhaled as radioactive carbon dioxide gas.

Like Oyama's experiment, Levin's experiment immediately produced results. The Geiger counter registered 500 counts per minute from the background. That is, 500 counts per minute was the signal level from the Geiger counter without any biological activity taking place in the soil. Shockingly, 9 hours after the experiment started, the Geiger counter was registering 4,500 counts per minute. After one day, the count rate had climbed to about 10,000 counts per minute, at which time it leveled off. Was biological material inhaling Levin's nutrients and exhaling radioactive carbon dioxide?

These early measurements led to the July 31 press conference at which Klein mentioned that the detections of oxygen in Oyama's Gas-Exchange experiment and carbon dioxide in Levin's Label-Release experiment might be mimicking biological signals. Had the scientific teams waited a few more days, or even paid more attention to their own reservations that first day, they likely never would have even made those statements. Over the days that followed, they quickly recognized that nonbiological, chemical reactions could yield the results they observed. The Gas-Exchange and Label-Release experiments "confirmed the presence in the surface of very reactive, oxidizing species,"[11] by which they meant a kind of molecule that produced oxygen when water was put in contact with the soil or that produced carbon dioxide when water containing organic compounds was put in contact with the soil. Neither experiment required biological activity to generate the measured signals.

The third experiment, the Pyrolytic-Release experiment, also produced some early results that generated speculation about a possible biological origin of the experimental output. Professor of biology Norman Horowitz, a geneticist at the California Institute of Technology, had designed this experiment to perform tests under conditions similar to those that he argued actually exist on Mars. That is, his experiments were carried out without the addition of any water. Mars is dry, he said. Martian life, if it exists, must survive without water, and so his experiments were designed without water. Instead, he measured whether any putative Martian organisms living in the soil took in carbon monoxide or carbon dioxide from the air (for the experiment, Martian air was injected into the experimental apparatus) and formed carbon-containing substances. Horowitz provided radioactively tagged carbon monoxide and carbon dioxide molecules, in order to track the uptake of those molecules by any Martian organisms.

After the Martian organisms, if they existed, had had enough time—120 hours was the given window of opportunity for them to become active and metabolize the CO and CO_2—the soil in the chamber was to be heated to a temperature of $635\,^{\circ}C$ ($1,175\,^{\circ}F$). The heating was designed to force the more volatile gases out of the soil into the air, from which they could be detected. Horowitz's instrument would measure the radiation emitted by the radioactive isotopes of CO and CO_2 released after heating. These gaseous species would be a diagnostic of the amount of carbon ingested and metabolized by Martian life-forms.

The Pyrolytic-Release experiment did yield carbon-rich molecules, i.e., organic material. Ultimately, Horowitz and his team found, however, that the carbon-bearing material could not be destroyed when heated to extreme temperatures and therefore could not be biological. Instead, they concluded, these materials were produced by chemical reactions with iron-rich minerals that are naturally abundant in the Martian soil.

In a meeting just one hour before the first press conference at which Klein announced to the world "at least preliminary evidence for a very active surface material" that "looked very much like a biological signal," the science teams were already debating alternative explanations.[12] Those explanations included surface minerals called oxidants, including compounds known as oxides, peroxides, and superoxides. All of these compounds, in the right circumstances, give up oxygen atoms in chemical reactions and thus could have been the source of the excess oxygen detected in Oyama's experiment. Eventually, the overwhelming consensus of Oyama and his scientific colleagues was that the spontaneous release and initial buildup of oxygen levels, as measured in the Gas-Exchange experiment, were chemical reactions involving compounds such as hydrogen peroxide. Their initial instincts would be strongly supported by several more years of laboratory experiments. Nevertheless, the combination of their desire to discover life, their optimism for finding life on Mars, and the encouragement they received from NASA to present to the public an exciting and headline-capturing discovery may have influenced their judgment in choosing how to present their scientific results to the public at the first scientific press conference about the Viking lander biology experiments. If they had listened to their scientific instincts, the headlines might have been much less exciting, but they would never have been misleading.

organic compounds? no or yes

The Viking landers had one more instrument whose measurements were extremely relevant to the biology experiments, even though that instrument was not formally part of the biology package. Klaus Biemann, of the Massachusetts Institute of Technology, led the team that built a miniaturized gas chromatograph with a mass spectrometer (GCMS), whose purpose was to detect and identify organic (carbon-containing) compounds near the surface of Mars. The experiment worked by heating and vaporizing a tiny soil sample. The soil was heated gradually, first to $50\,^{\circ}$C, then to $200\,^{\circ}$C, $350\,^{\circ}$C, and $500\,^{\circ}$C. A stream of hydrogen gas, which was sprayed into the GCMS, carried any vaporized gases into a chamber, where the instrument measured the light emitted by the vapors and then measured the masses of the particles in the vapor.

Biemann and his science team reported, definitively, "No organic compounds were found at either of the two landing sites." No naphthalene ($C_{10}H_8$, which is the primary ingredient in mothballs), no benzene (C_6H_6, a manufactured derivative of coal and petroleum), no acetone (C_3H_6O, a naturally occurring material on Earth, formed during the metabolic breakdown of animal fat), no toluene (C_7H_8, first distilled from pinesap in 1837 and used as a solvent, for example in paint thinner). These experiments demonstrated that nothing in the Martian soil is effective at synthesizing organic molecules. In addition, since carbon-bearing meteoritic dust should steadily filter down through the atmosphere, the gas chromatograph/mass spectrometer experiments demonstrated that the Martian soil (or interactions of ultraviolet light and cosmic rays with the Martian soil) is effective at destroying any of those molecules that rain down onto the surface of Mars. This experiment was definitive: the absence of organic compounds in these experiments "rules out carbonaceous remnants of now extinct biological or abiological processes in the surface material of these two sites."[13]

The GCMS experiment was critical to understanding and interpreting the results of the Viking biology experiments. If the GCMS experiment had been sent to Mars on a mission prior to the Viking missions, the Viking biology experiments would never have been proposed, let alone designed, built, and sent to Mars. According to Klein, Horowitz, and Biemann, writing in 1992, "The absence of organic compounds at these two very distant (from each other) sites demonstrated that there is presently neither biological nor abiological synthesis of organic compounds occurring . . . What became clear even during the Viking mission was that if the GCMS results were correct (and there was reason to believe that this was the case) the three biological experiments had essentially lost their original purpose. With no detectable trace of organic matter in the surface material, there was no possibility of finding extant life at the two landing sites."

Furthermore, they wrote, "The Viking findings established that there is no life at the two landing sites, Chryse and Utopia. Although the two sites are 25° apart in latitude and on opposite sides of the planet, they were found to be very similar in their surface chemistry. This similarity reflects the influence of global forces such as extreme dryness, low atmospheric pressure, short-wavelength ultraviolet flux, and planet-wide dust storms in shaping the Martian environment. These same forces virtually guarantee that the Martian surface is lifeless everywhere."[14] Strong stuff.

The conclusion that the two Viking landers found no evidence of life on Mars is the overwhelming conclusion of the scientific community; that conclusion, however, is not quite unanimous. In fact, one senior Viking team leader, Gilbert Levin, has been voicing his opinion for almost four decades that "At both landing sites, some 4,000 miles apart, the LR [the "Labeled Release" experiment] returned evidence of living microorganisms. Initially discounted by NASA and most space scientists," he wrote, "the results of this milestone project have, nonetheless, been causing excitement and controversy ever since." Writing about himself and his own work in the third person, Levin summarized his scientific point of view in 2015 on his personal website as follows: "In 1997, after 21 years of study of the Mars LR results, of new information scientists obtained about environmental conditions on Mars, and of the extreme environments in which life was found on Earth, Dr. Levin published his conclusion that the LR had, indeed, discovered living microorganisms on the Red Planet."[15]

Levin and his colleague Patricia Straat, in 1976 also of Biospherics, Inc., have consistently maintained that the Label-Release experiment indicates the presence of organisms in the soil samples studied by Viking. Writing in 1979, in *Science*, they concluded, "despite all hypotheses to the contrary, the distinct possibility remains that biological activity has been observed on Mars."[16] A decade later, Levin and Straat (at that time working at the National Institutes of Health), wrote:

A decade has passed since the first labeled-release (LR) Viking biology experiment produced an astonishing positive response on Mars. But that response was deemed unconvincing when no organic compounds was [*sic*] found. As a result, many attempts have been made to explain the LR data without invoking life. The dominant theory expounded hydrogen peroxide as a chemical agent, suggesting that it reacted with one of the nutrient compounds to mimic a biological response. This theory was tested and essentially disproved on Mars. There is in fact no evidence that it exists on Mars, and even if it formed it would be destroyed by the environment long before it could affect an experiment. We have carefully tested all of the nonbiology theories and have found none to be scientifically adequate. We also verified that the GCMS organic detection sensitivity may have missed very low densities of organic matter. It is now our contention that the survival of the LR data, together with other information not previously considered (including Viking

Plate 1. Living marine stromatolites in Hamlin Pool, Australia. Image courtesy of Kristina D. C. Hoepper/Creative Commons at https://www.flickr.com/photos/4nitsirk/11902636365

Plate 2. Hubble Space Telescope image of Mars, as seen in 2001 at a distance of 43 million miles (68 million kilometers) from Earth. Ice is evident at the southern polar cap (bottom) while a dust storm obscures the northern polar cap (top). A second giant dust storm can be seen in the Hellas Basin (lower right). Water ice clouds are seen surrounding the north polar cap, extending northward from the south polar cap, and near the Martian equator. Image courtesy of NASA and the Hubble Heritage Team (STScI/AURA).

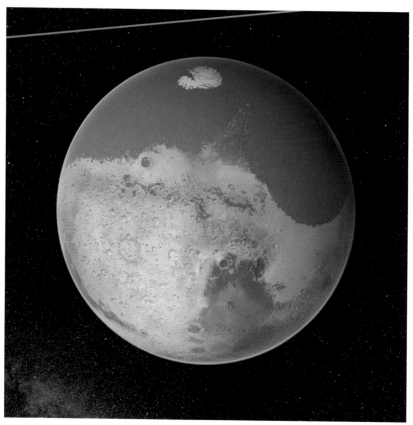

Plate 3. Conceptual map showing a large ocean, holding more water than Earth's Arctic Ocean, on ancient Mars, as reported by Villanueva et al. in *Science* in 2015. Given current Martian topography, which likely has not changed significantly in billions of years, the ocean would be in the low-elevation Northern Plains region. The idea that Mars once had an ocean emerges from measurements that indicate Mars has lost 85 percent of the water it once had to space. Note that if ocean basins once existed on Mars, they would have filled enormous, ancient impact craters, whereas ocean basins on Earth are the result of plate tectonics, which is and has always been absent on Mars. Image courtesy of NASA's Goddard Space Flight Center.

Plate 4. Catherine King-Frazier, a member of the 1984–1985 National Science Foundation–sponsored research team tasked to search for meteorites in Antarctica, walking across the Allan Hills Main Ice Field on a blustery day in early December 1984. The meteorite ALH 84001 was found by meteorite search team member Roberta Score later that month about 50–60 kilometers from this location, in the Allan Hills Far Western Ice Field. Image courtesy of Roberta Score.

Plate 5. Martian meteorite ALH 84001, photographed in the laboratory at Johnson Space Center. For a sense of scale, the small black cube (lower right) is 1 cm (four-tenths of an inch) on a side. Parts of the outside of the meteorite are covered with a black fusion crust, while the inside is greenish gray. Image courtesy of NASA's Johnson Space Center.

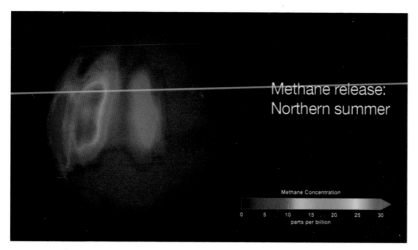

Plate 6. Map of multiple plumes of methane, seen in Martian summer in the northern hemisphere in 2003, as reported by Mumma et al. (*Science*, 2009). The methane appears to be present globally, but is strongest in limited locations. Image courtesy of NASA.

Plate 7. Mars Global Surveyor spacecraft, as imagined in orbit above Mars. Image courtesy of NASA/JPL-Caltech.

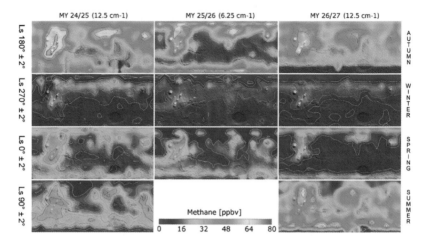

Plate 8. Maps of the spatial distribution of the apparent abundance of methane on Mars, showing differences in methane abundance with longitude and season, as calculated from Mars Global Surveyor data. Each map extends in latitude from –60 degrees (south) to +60 degrees (north). In each Martian year (vertical columns for MY 24/25, MY 25/26 and MY 26/27) maps are shown for each Martian season. For each Martian season, a longitudinal slice is presented for each of the three years. For Martian Autumn, the longitude near 180 degrees is shown; for Martian Winter, the longitude centered on 270 degrees is shown; for Martian Spring, 0 degrees, for Martian Summer, 90 degrees. Colors indicate the abundance of methane, from a high of about 80 parts per billion (red) to a low of a few parts per billion (blue). At all seasons, more methane is seen at latitudes of 90 and 180 degrees than at 0 and 270 degrees. And more methane is seen in Summer and Autumn than in Winter and Spring in all three Martian years. Image from Fonti and Marzo, *Astronomy & Astrophysics* 2010; Reproduced with permission from *Astronomy & Astrophysics*, © ESO.

Plate 9. Mars Curiosity Rover self-portrait, obtained on May 11, 2016, at a drilled sample site called "Okoruso," on the "Naukluft Plateau" of lower Mount Sharp. An upper portion of Mount Sharp is prominent on the horizon. For scale, the rover's wheels are 20 inches (50 centimeters) in diameter and about 16 inches (40 centimeters) wide. Image courtesy of NASA/JPL.

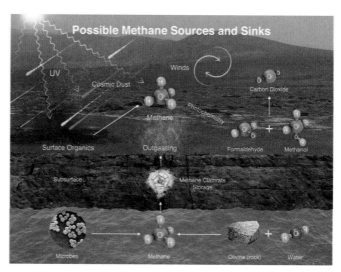

Plate 10. Illustration of several possible ways that methane might be added to and removed from the Martian atmosphere. Subsurface sources include methanogenic microbes, serpentinization (water reacting with olivine minerals), and ancient methane escaping from clathrates. Surface sources include ultraviolet light reacting with carbon-rich dust that rains down onto the surface. Mechanisms that might destroy methane include surface chemical reactions that utilize oxygen-rich molecules to process methane into carbon dioxide and swirling dust devils that generate fast-moving electrons that act to destroy the methane. Image courtesy of NASA's Goddard Space Flight Center/Brian Monroe.

Lander image and spectral data that suggest the possible existence of Martian lichen), justifies the conclusion that it is now more probable than not that the LR experiment did in fact detect life on Mars.[17]

Although Levin's interpretation of the Viking data is an extreme outlier, some additional support for Levin's point of view emerged in 2010 when Rafael Navarro-Gonzalez, of the National Autonomous University of Mexico, and Chris McKay, of NASA's Ames Research Center in California, performed an experiment on a handful of dirt taken from the extremely high (13,000 feet or 4,000 meters) and dry Atacama Desert in Chile. After dosing the dirt with perchlorate (the negative ion ClO_4^-) salt, a very reactive, oxidizing chemical containing only chlorine and oxygen, and heating the sample, they detected the presence of two compounds—chloromethane (CH_3Cl) and dichloromethane (CH_2Cl_2)—that formed when the perchlorate reacted with and consumed other materials in the dirt.[18] Both of these newly formed compounds contain both carbon and hydrogen atoms not present in the perchlorate. The chlorine atoms obviously came from the perchlorate. Where did the carbon come from? The carbon atoms must have come from carbon-bearing molecules that were already present in the soil. That is, the formation of chloromethane and dichloromethane is indirect proof of the presence of organic material in the Atacama Desert dirt. We need to be cautious here. Nearly all, but not all, carbon-bearing molecules are, by definition, organic molecules. (An organic molecule must contain carbon, but not all molecules with carbon are considered organic. To be organic, the carbon-containing compounds must contain carbon bound to hydrogen. SiC (silicon carbide), WC (tungsten carbide), and CO (carbon monoxide) are examples of non-organic, carbon-containing molecules.) But organic molecules are not necessarily evidence for the presence of life, as organic molecules can exist in any environment in which carbon and hydrogen are present. Furthermore, the carbon could have come from atmospheric carbon dioxide or carbon monoxide adsorbed onto minerals in the Atacama Desert dirt.

The connection to Mars and the Viking biology experiments is the following: in 1976, Viking 1 detected chloromethane at a level of 15 parts per billion and Viking 2 detected dichloromethane at levels of between 2 and 40 parts per billion. These are the same two chemicals detected by Navarro-Gonzalez and McKay in their experiment with the Atacama Desert dirt. The Viking teams attributed the detections of both the chloromethane and the dichloromethane to terrestrial contaminants.

Planetary scientists make extraordinary efforts to reduce the amount of terrestrial gases that seep into their detector chambers and hoses before launch, but they cannot completely eliminate molecules from Earth's atmosphere from traveling to Mars as stowaways. The two Viking landers brought these two compounds with them from Earth, said the Viking scientists. Navarro-Gonzalez and McKay, however, now think their laboratory work with Atacama Desert dirt proves otherwise.

Levin gained one more possible line of support after NASA's Phoenix mission landed in the far north of Mars on May 25, 2008, where it operated for about five months before succumbing to the Martian winter. Phoenix used its robotic arm to scrape up some icy dirt and subject that dirt to onboard testing. One of the discoveries made by Phoenix is that perchlorates are present in the dissolved salts in the Martian soil at a level of 0.4 to 0.6 percent.[19] From an astrobiology point of view, this discovery is extremely significant when combined with the Atacama Desert experiment. Putting these pieces together, Navarro-Gonzalez and McKay argue that the chloromethane and dichloromethane detected by Viking could have been due to pure Martian chemistry. Martian perchlorates could have reacted with Martian organic molecules in the soil to produce the chloromethane and dichloromethane detected by Viking 1 and Viking 2. The conclusions previously drawn from the GCMS experiment, they contend, might be wrong. In this interpretation of the Viking results, the Viking soil samples must have had some organics in them. If so, they argue, the results of Levin's Label-Release experiment might be proof that life exists on Mars. The presence of organic molecules in Martian soil does not prove that life exists or existed on Mars, but Chris McKay, at least, connects the dots from organic molecules to biology: "There's a possibility that some of those organic molecules are in fact biomarkers," he wrote.[20]

11

---

hot potato

During the last days of 1984, Roberta Score, the curator of the Antarctic Meteorite Laboratory at the Johnson Space Center in Houston, Texas, was hunting for meteorites in Antarctica. She had been hired by NASA in 1978, fresh out of UCLA where she had majored in geology, to help get this Meteorite Laboratory up and running. In the late 1960s and early 1970s, NASA had learned how to carefully curate and store 840 pounds of rocks and dirt collected by Apollo astronauts on the surface of the Moon. Now, in support of the National Science Foundation's Antarctic Search for Meteorites (ANSMET) program, the lab was adding to the moon rock collection more extraterrestrial rocks: pristine meteorites collected on the blue ice of Antarctica that had fallen thousands or millions of years ago, then been buried by snowfall, preserved in ice, and finally unburied by the motion of glaciers and the scouring action of the wind.

Score and the other members of the 1984–1985 collection team, John Schutt, Carl Thompson, Scott Sandford, Bob Walker, and Catherine King-Frazier, lived in tents in the remote Antarctic cold; they rode snowmobiles across dangerous terrain, working diligently to avoid plunging into crevasses or acquiring frostbitten fingers and noses. Robbie Score learned, quickly, that scouring the Antarctic in search of meteorites is an extreme sport, not child's play, but for the scientific community these meteorites were a scientific windfall, well worth the difficult work involved in obtaining them.

chapter 11

Figure 11.1. Catherine King-Frazier, a member of the 1984–1985 National Science Foundation–sponsored research team tasked to search for meteorites in Antarctica, walking across the Allan Hills Main Ice Field on a blustery day in early December 1984. The meteorite ALH 84001 was found by meteorite search team member Roberta Score later that month about 50–60 kilometers from this location, in the Allan Hills Far Western Ice Field. See Plate 4. Image courtesy of Roberta Score.

On the twenty-seventh of December, Score got lucky when she found an odd-looking, greenish-gray-colored, four-and-a-quarter-pound rock that she correctly surmised was a meteorite. When she lifted that potato-shaped rock off the Antarctic ice, Earth shifted, figuratively at least. She released a shock wave that grew in energy until it triggered an explosion of ideas a dozen years later. That later explosion sent ripples through the staid worlds of astronomy, biology, geology, geophysics, and planetary sciences that dramatically reshaped all of those disciplines.

On that bitter cold Antarctic morning, albeit a morning after a night during which the Sun never set, Robbie Score was careful not to contaminate the meteorite by touching it with her bare hands. With help from the rest of the collection team, she put this unusual looking, black-encrusted rock into a NASA-provided "clean bag" and sealed the bag with tape. Months later, back in Houston, she had the responsibility of assigning numbers to each meteorite she and her

Figure 11.2. Martian meteorite ALH 84001, photographed in the laboratory at Johnson Space Center. For a sense of scale, the small black cube (lower right) is 1 cm (four-tenths of an inch) on a side. Parts of the outside of the meteorite are covered with a black fusion crust, while the inside is greenish gray. See Plate 5. Image courtesy of NASA's Johnson Space Center.

teammates had collected in Antarctica. Score chose to pull out her favorite meteorite first. It was her favorite because it was green, and meteorites are not typically green, though she was surprised to find that in the light of the laboratory environment it looked mostly gray. She tagged it and then numbered it 001. Score's green rock was not the first one found in the 1984–1985 Antarctic season of meteorite hunting, but it fittingly became the first one tagged upon her return to her day job in Houston. In doing so, she gave a name to the meteorite that a decade hence would compete with the Rosetta Stone for the title of the most famous rock on Earth. Robbie Score's rock, collected in the Allan Hills section of Antarctica in 1984, is now known as ALH 84001, and in her honor the section of the Meteorite Hills region of the Darwin Mountains where she found this rock that, thousands of years ago, fell from the sky is now named Score Ridge.[1]

As was done with all meteorites that were entered into the NASA Antarctic meteorite collection at Johnson Space Center, a technician carefully sliced off a half-gram chip of ALH 84001 and shipped the rock fragment to the Smithsonian Museum of Natural History, in Washington, DC. At the Smithsonian, a meteorite specialist carried out a quick examination of this slice of ALH 84001 and then assigned the parent meteorite to a classification category within the known framework of meteorite families. This cursory examination is performed for every meteorite collected in Antarctica, and the subsequent publication of those write-ups in the *Antarctic Meteorite Newsletter* allows meteoriticists from around the world to decide whether any of the new arrivals are worthy of further study in their laboratories.

Because ALH 84001 was numbered 001 by Robbie Score and because its greenish color was so unusual, this rock was a high-priority meteorite when it arrived in Houston, and the small chip of ALH 84001 was a high-priority sample for Glenn MacPherson when it showed up in his lab at the Smithsonian. However, MacPherson, the curator who examined the small slice of ALH 84001, quickly decided it was not a particularly interesting object. He did, however, discover the first important clue that, one day, would lead to ALH 84001 becoming a worldwide phenomenon: he correctly identified the rock as igneous, or volcanic, in origin.

ALH 84001, like almost all small rocks, is a chip off an old block. The old block, the larger "parent" rock of which ALH 84001 is a fragment, must have formed in an environment in which the temperature was high enough that the precursor materials that formed the rock were molten. Magma (molten or semi-molten material that remains below the surface of a planet or moon or asteroid) and lava (magma that erupts onto the surface) can only exist on objects in the solar system that are as large or larger than the biggest asteroids. Only large objects can both generate enough heat from the decay of radioactive minerals, like certain isotopes of aluminum, potassium, thorium, and uranium, and then trap that heat within them for long enough for their internal temperatures to rise to a level that produces volcanic activity on their surfaces or magmatic activity deep within. Meteorites that formed as igneous rocks that cooled slowly, deep inside a large parent-body, are called diogenites. ALH 84001 was born as a diogenite in such an environment.

In 1984, the object considered the principal source of all known diogenites was Vesta. Vesta is a very large asteroid (about 525 kilometers, or 326 miles, in diameter), second in size only to Ceres among objects in the asteroid belt. Because of Vesta's large size, planetary scientists are certain that soon after it formed, early in the history of the solar system, radioactive materials buried deep inside of it, probably mostly the unstable isotope of aluminum known as aluminum-26, decayed, released heat, and melted the inside of Vesta. As a result, reduced iron, which is the element iron that has been chemically isolated from oxygen, drizzled downward through the magma until it reached Vesta's center and formed an iron-rich, oxygen-poor core. In addition, as gravity pulled the iron downward through the softened innards of Vesta, the iron dragged with it other iron-loving (siderophile) elements. At the same time, lighter, rocky (lithophilic) elements bubbled up toward Vesta's surface. The basaltic lavas that rose up and then solidified to form the mantle and outer crust of Vesta are presumed to be the source for igneous meteorites like ALH 84001.

Of course, objects in the solar system that are larger than Vesta, for instance Earth's Moon, or even the Earthlike planets, Mercury, Venus, and Mars, are also places where internal melting took place, convection brought lighter magmas upward, and surface volcanic activity occurred. Therefore, they also could be sources of basaltic meteorites, although some were recognized as less likely potential sources of meteorites than others.

In order for a meteorite to come from a large solar system body, several things must happen. First, an asteroid of significant size must hit the surface of that moon or planet and some of the debris from that impact must be lofted off the surface intact and at high speed. The impact debris kicked off the surface then must drill a hole through the planet's atmosphere, if the planet has one, and emerge above the atmosphere with a high enough velocity (known as "escape velocity") to escape the gravitational clutches of the planet.

Venus has a massive atmosphere and an escape velocity (6.34 miles per second, or 10.4 kilometers per second) almost as large as that of Earth (6.96 miles per second, or 11.2 kilometers per second). The likelihood of any meteorites being splattered off the surface of Venus, pushing through Venus's dense atmosphere, emerging with a velocity greater than escape velocity, and getting to Earth is thought to be nearly zero. At this time, we know of no meteorites from Venus in our collections.

Mercury has no atmosphere and an escape velocity (2.7 miles per second, or 4.3 kilometers per second) that is fairly small compared to that of Venus; but even though Mercury lacks a substantive atmosphere, to get from Mercury to Earth a meteorite also would have to battle against the Sun's gravitational pull to move outward away from the Sun from Mercury's orbit to Earth's orbit. Rising 57 million miles upward from a place near the Sun in order to reach Earth is a serious energy challenge. As with Venus, we know of no meteorites from Mercury in our collections. (One 4.5-billion-year-old meteorite, NWA 7325, which was found in southern Morocco in 2012, might be from Mercury, thought that claim has been challenged.[2]) As a result, the planetary science community, in 1984, considered neither Mercury nor Venus seriously as a likely candidate source for igneous meteorites.

On the other hand, both the Moon (escape velocity only 1.5 miles per second, or 2.4 kilometers per second) and Mars (escape velocity of 3.1 miles per second, or 5.0 kilometers per second) were considered plausible meteorite sources because of the combination of their lower escape velocities, their absent (Moon) or thin (Mars) atmospheres, and their relative proximity to Earth; however, in the early 1980s the absence of any actual meteorites in meteorite collections from either object suggested otherwise.

A meteoritic breakthrough occurred on January 18, 1982, when John Schutt, of Spokane, Washington, the leader of the 1981–1982 U.S. search party looking for meteorites in Antarctica (and the leader of Robbie Score's team three years later), found an unusual-looking, 31-gram (about 1 ounce) rock now known as ALH 81005, which showed similarities to lunar rocks. By 1983, several teams of meteoriticists, working independently, had confirmed that this specimen was, without any doubt, a lunar meteorite.

Finally, the planetary science community had confirmed that meteorites could escape the gravitational grip of the Moon and survive a trip to Earth. Nevertheless, since ALH 84001 did not bear any resemblance to Moon rocks, as did ALH 81005, and since the evidence on the ground in the form of actual meteorites seemed to demonstrate that meteorites either cannot or do not travel from Mars to Earth, in 1984 Vesta remained the most likely source for all diogenites.

Vesta had another very significant plus on its side of the ledger. The spectrum of sunlight reflected from the surface of Vesta and observed by astronomers appears very similar to the results of laboratory experiments in which meteoriticists study the spectra of light reflected from meteorites known to have volcanic origins. In contrast, the reflected light spectra from other kinds of meteorites do not mimic the spectrum of Vesta. In addition, the reflected light spectra of other asteroids do not match the spectra of the diogenites or of Vesta. Vesta and these meteorites were a unique match: no other meteorites are a good match to Vesta, and no other asteroids are a good match to these meteorites. All of these experimental results strongly suggested that Vesta was the likely source of all meteorites with igneous origins. This result was not controversial; rather, it was part of the broad consensus as to how the planetary science community understood the rocks in our meteorite collections.

When the tiny rock fragment chipped off ALH 84001 arrived at the Smithsonian, it did not present a particularly difficult challenge for MacPherson. With ease, he immediately classified ALH 84001 as a diogenite. As a diogenite, ALH 84001 is nothing special, just one more meteorite among the nearly 1,000 such objects thought to be from Vesta.*

ALH 84001 remained unbothered and unstudied until 1988, at which time David Mittlefehldt, a geochemist from Lockheed Engineering working for NASA in Houston, began a study of meteorites presumed to be from Vesta. Mittlefehldt used a device called an electron microprobe, which bombards a microscopic sample under study with a beam of electrons. This is a nondestructive technique that causes the elements in the sample to emit X-rays. The energies of the emitted X-rays are diagnostic of the elements present in the material under study.

By 1990, Mittlefehldt had determined with a high degree of confidence that the X-rays produced by ALH 84001, when subjected to his

*The Meteoritical Society identifies 383 meteorites as diogenites (basaltic meteorites composed mostly of the mineral orthopyroxene). In addition, 279 meteorites known as eucrites (basaltic meteorites composed mostly of the mineral pigeonite) and 329 meteorites known as howardites (meteorites containing a mixture of diogenite and eucrite material) are also thought to be from Vesta.

probing, did not match the X-ray signature produced by other meteorite fragments presumed also to be from Vesta. Clearly, in ways that were obvious but that he did not understand at that time, ALH 84001 was different in some fundamental way from all other known diogenites. After three more years of struggling with ideas and searching for a plausible explanation of this discrepancy, he figured out part of the answer. He had to challenge and then toss the assumption that ALH 84001 was a fragment of Vesta. Yes, the other diogenites are all from Vesta, but ALH 84001 is not. Of course, knowing that ALH 84001 is not from Vesta simply created a new mystery: which other solar system body, one large enough to generate magma and slowly cool that material deep inside, was the birthplace of ALH 84001?

The X-ray spectrum of ALH 84001, which told him the atomic contents of the rock, is equivalent to a set of fingerprints. In this case, Mittlefehldt relied on the published work of other meteoriticists who had previously measured the X-ray fingerprints of some other unusual meteorites. He found that the fingerprints of ALH 84001 did match the known atomic fingerprints of a very small handful of other meteorites, and those other meteorites were not diogenites from Vesta. The remarkable thing about the match was that all of the other meteorites whose X-ray signatures matched that of ALH 84001 were rocks from Mars. Mystery solved: ALH 84001 did not come from Vesta. It came from Mars[3]; however, it was different from the other known Martian meteorites.

In 1993, Robert N. Clayton, a highly respected meteoriticist at the University of Chicago, confirmed the Martian origin of this meteorite by analyzing the isotopes of oxygen in the rock.[4] Clayton, in 1973, had discovered that the ratios of the three stable isotopes of oxygen—oxygen-16, oxygen-17, and oxygen-18—were of enormous significance for cosmochemistry. That is, they could be used to identify the location in the solar system from which a rock sample originated and could even be used to identify the distinct events that created the oxygen isotopes before the Sun and solar system formed. Clayton was elected to the Royal Society of Canada in 1980, the Royal Society of London in 1981, and the National Academy of Sciences in 1996. Confirmation by Robert Clayton of the Martian origin of ALH 84001, based on the oxygen isotope ratios in the meteorite, was akin to being blessed by the pope for a member of the Roman Catholic faith.

What are these isotopes of oxygen? The oxygen atom always has eight positively charged protons in the atomic nucleus and usually also commonly has eight uncharged neutrons in the nucleus. Such an atom is known as oxygen-16 (^{16}O). But stable atoms of oxygen can also have nine neutrons (^{17}O) or 10 neutrons (^{18}O) sharing the nucleus with the eight protons. The relative percentages of ^{16}O atoms (99.76 percent) to ^{17}O atoms (0.039 percent) and ^{18}O atoms (0.201 percent) on Earth (the so-called Standard Mean Ocean Water, or SMOW, abundances) are well known. In addition, the fact that these relative abundances can change, or fractionate, as a result of mass-dependent processes like evaporation (what remains in liquid form is enriched in the heavier isotopes because of the effect of gravity pulling downward more strongly on the heavier isotopes) or chemical bonding (heavier isotopes are bound more strongly than lighter isotopes) is well understood. Any sample of water from any environment on Earth will yield deviations from the SMOW abundances that are directly related to the original SMOW values, which were determined by the isotopic abundances that were present in the primordial reservoir of materials out of which Earth formed. When these deviations are plotted on a graph that depicts the deviations in the ^{18}O values versus the deviations in the ^{17}O values, every drop of terrestrial water will plot on a single line known as the terrestrial fractionation line. This terrestrial fractionation line is identical for both Earth and the Moon, which tells us that they formed from the same original reservoir of material. Other solar system bodies, however, whether Mars or Jupiter or different types of meteorites, have their own, unique oxygen isotope fractionation lines. These isotopic differences tell us that each planet formed from materials with unique isotopic ratios. Once known and understood, these isotopic ratios become a signature of origin for meteorites and for planets. They are, effectively, fingerprints that identify the home planet (or moon or asteroid) of a rock.

In 1985, a year after ALH 84001 descended into obscurity as just another diogenite, and at a time when no planetary scientist had yet identified even a single meteorite from another planet, Robert

Pepin, of the University of Minnesota, published a remarkable research result that revealed a one-to-one correlation between the relative amounts of rare gases in the atmosphere of Mars, as had been originally measured by the Viking lander in 1976,[5] and the relative amounts of those same gases trapped in air bubbles in the meteorite known as EETA 79001.[6] EETA 79001, like ALH 84001, was an Antarctic meteorite, collected in 1979 in the Elephant Moraine region. Pepin found that the relative amounts of xenon-132, krypton-84, argon-36, argon-40, neon-20, molecular nitrogen (N_2), and carbon dioxide in the air bubbles in EETA 79001 were indistinguishable from, in fact a perfect match to, the gases found by Viking in the atmosphere of Mars. Without any doubt, EETA 79001 was a piece of Mars.

How did it get knocked off of Mars, travel through the solar system, and finally land on Earth, in Antarctica, where it sat for thousands of years, quietly waiting to be collected and tagged by an Antarctic meteorite collection team? Those are questions worth asking, but for ALH 84001, the critical piece of information is not how EETA 79001 left Mars and traveled to Earth. No, the critical data point is that when EETA 79001 was knocked off of Mars, it trapped some Martian atmospheric gases in air pockets within the meteorite, protected those air bubbles for millions of years, and successfully navigated the Mars-to-Earth trip. As one last contribution to our understanding of other meteorites, EETA 79001 provides the instruction set for testing whether other meteorites are from Mars: test the gases trapped in air bubbles in a particular meteorite and see how they compare to the Martian atmosphere.

In the case of ALH 84001, Robbie Score's rock did have tiny bubbles of gas trapped within it, and the isotopic signatures of the atoms in those bubbles of trapped gas were identical to those of the gases in the Martian atmosphere. Bingo. ALH 84001 was now known to be similar in this way to a handful of meteorites already known to be from Mars, because they also had tiny bubbles of Martian gas trapped within them. Thus, the Martian origin of ALH 84001 was firmly established and remains uncontroversial.

Even among the small family of Martian meteorites, however, ALH 84001 stood apart. The small number of known Martian meteorites include the Shergottites, the Nakhlites, and the Chassignites (collectively known as SNCs). The Shergottites are named for the town of Shergotty (now Sherghati), in northeastern India, where the Shergotty meteorites fell in 1865. More than one hundred Shergottites have been identified, collected from Antarctica, California, Libya, Algeria, Tunisia, Mali, Mauritania, Nigeria, and Oman. The largest is just over 8.5 kilograms (about 19 pounds), the smallest only 4.2 grams (0.15 ounces). The Nakhlites are named for the village of Nakhla, a tobacco-farming village about 25 miles (40 kilometers) southeast of Alexandria, Egypt, where these meteorites fell on June 28, 1911. The Nakhlites, composed of forty individual fragments, were originally found after a farmer reported seeing smoke trails and explosions in the sky and claimed that he witnessed a dog that was killed by a falling rock fragment. Although the Egyptian Survey Department, after later interviews, concluded that the dog story was likely imagined, the Nakhla dog remains one of only a very few living things (possibly) struck and killed by meteorites.* A total of eighteen Nakhlites are known, including the Lafayette meteorite, found in Indiana in 1931; the Governador Valadares meteorite, found in Brazil in 1958; and a handful of others found in the last two decades in Antarctica, Morocco, and Mauritania. The largest Nakhlite is 13.71 kilograms (30.2 pounds); the smallest is only 7.2 grams (0.25 ounces). And the Chassignites are named for the hamlet in northeastern France called Chassigny, where these meteorites fell in 1815. Only two other Chassignites have been identified, both small rocks found in northwest Africa.

All of the SNCs formed from a molten reservoir on the surface of Mars about 1.3 billion years ago. In stark contrast, ALH 84001 formed *inside* Mars 4.1 billion years ago[7] and is nearly as old as Mars. ALH 84001 formed almost as soon as Mars was cool enough to allow any rocks to form on that planet or on any planet in the solar system. Suddenly, as the oldest known rock in the solar system from a planet—a few meteorites known as carbonaceous chondrites, which

---

*Reported deaths (not all with supporting evidence) from meteorites include a man in India in 1825, cattle in Brazil in 1836, a horse in Ohio in 1860, a whole family in China in 1907 (no evidence), the Nakhla dog in Egypt in 1911, a man in a wedding party in Yugoslavia in 1929, and hundreds of reindeer, among other creatures, by the Tunguska blast in Siberia in 1908 (see www.icq.eps.harvard.edu/meteorites-1.html).

are not from planets, are just a bit older—plus as a rock from Mars, ALH 84001 became a hot potato, with meteorite specialists from all over the world asking for a piece of it to study.

Fossils from mars

On August 7, 1996, ALH 84001 became the most interesting and controversial rock in the world. On that day, David McKay, Everett Gibson, Jr., and Kathie Thomas-Keprta, all highly respected career scientists based at the Johnson Space Center in Houston, and professor of chemistry and of physics Richard Zare, a distinguished laser chemist at Stanford University and a member of the National Academy of Sciences, represented their team of nine co-authors at a NASA-sponsored press conference in Washington, DC. They announced that they were publishing a paper in *Science*, the contents of which NASA thought worthy of having members of this team present to the world at a press conference hosted by NASA. In that paper, they were putting forth the claim that they had discovered evidence inside ALH 84001 of fossils that strongly suggested that life had existed on Mars in the ancient past.[8] "Although inorganic formation is possible," they wrote, "formation of the globules by biogenic processes could explain many of the observed features, including the PAHs.* The PAHs, the carbonate globules, and their associated secondary mineral phases and textures could thus be fossil remains of a past Martian biota."

Unlike the observational evidence for canals, chlorophyll, lichens, or the Sinton bands, which, whether right or wrong, were only indirect evidence of life on Mars, at the press conference and in their published research paper McKay's team was showing pictures—pictures!—of objects they claimed were fossils of ancient Martian life-forms. No hyperbole is needed: if this discovery is correct, it is absolutely phenomenal, the discovery of the century.

Assuming the evidence reported by McKay, Gibson, Thomas-Keprta, and Zare is proven correct, how did life arise on Mars? Life

---

*PAHs are defined and discussed later in this chapter.

may have emerged independently on both Earth and Mars. Or, life may have arisen first on Mars and then been transferred to Earth via an asteroid collision with Mars. That collision may have launched a life-bearing meteorite into orbit around the Sun, which millions of years later fell to Earth!

The discovery reported by McKay and his team was important enough to warrant special remarks by President Bill Clinton, carefully orchestrated and coordinated by NASA. Speaking from the South Lawn of the White House, just before the NASA press conference began, Clinton told the world:

> It is well worth contemplating how we reached this moment of discovery. More than 4 billion years ago this piece of rock was formed as a part of the original crust of Mars. After billions of years it broke from the surface and began a 16-million-year journey through space that would end here on Earth. It arrived in a meteor shower 13,000 years ago. And in 1984 an American scientist on an annual U.S. government mission to search for meteors on Antarctica picked it up and took it to be studied. Appropriately, it was the first rock to be picked up that year—rock ALH 84001. Today, ALH 84001 speaks to us across all those billions of years and millions of miles. It speaks of the possibility of life. If this discovery is confirmed, it will surely be one of the most stunning insights into our universe that science has ever uncovered. Its implications are as far-reaching and awe-inspiring as can be imagined.[9]

President Clinton's remarks included a few mistakes: American scientists in Antarctica were looking for meteorites, not meteors. And ALH 84001 was the first rock from the 1984 collection mission to be catalogued, but not the first to be collected. But he conveyed the awe and wonder and importance of this discovery extremely well to a national audience.

McKay's discovery birthed NASA's Astrobiology Institute, which has grown over the decades since the McKay press conference; it inspired aggressive searches for extreme life-forms, now referred to as extremophiles, on Earth; and it initiated a vigorous public and scientific debate about what life is. This discovery also invigorated NASA with the support of the public and the U.S. Congress that has continued for several decades. As a result, NASA has sent a mission to Mars, and sometimes multiple missions, almost every other year for more than two decades. These missions have included two rovers that have finished their missions (Mars Pathfinder Rover 1996–1997; Mars Exploration Rover Spirit 2003–2011), two rovers that continue to work (Mars Exploration Rover Opportunity 2003–present; Mars Science Laboratory Rover Curiosity 2011–present), and another rover scheduled for launch in 2020 (the Mars 2020 Rover). NASA has also sent two landers to Mars: the Mars Polar Lander (1999) failed while the Mars Phoenix Lander (2007–2008) was a great success. A third, the InSight (Interior Exploration using Seismic Investigations, Geodesy and Heat Transport) Mars Lander, is scheduled for launch in spring 2018 and due to land on Mars that November. Finally, since 1990, NASA has put five spacecraft in orbit around Mars and sent a sixth orbiter (Mars Climate Orbiter, 1998–1999) that was lost upon arrival at Mars. The five successes include two completed missions—Mars Observer (1992–1993) and Mars Global Surveyor (1996–2006)—and three missions that remain active—Mars Odyssey (2001–present), Mars Reconnaissance Orbiter (2005–present), and Mars Atmospheric and Volatile Evolution (MAVEN, 2013–present).* In addition, teams of scientists and engineers are planning sample-return missions and for human exploration of the red planet in the not-too-distant future.

To their credit, McKay, Gibson, Thomas-Keprta, Zare, and their colleagues and co-authors Hojatolla Vali, Christopher Romanek, Simon Clemett, Xavier Chillier, and Claude Maechling followed established protocols when they published their scientific results. They submitted their research paper to *Science* before they announced their results to the press. The editors of *Science* subject every submitted manuscript to a refereeing process by other experts, known as peer review. The

---

*Three other spacecraft are active in Mars orbit. The Mars Express Orbiter (2003–present) was launched and is managed solely by ESA. In addition, the ExoMars Trace Gas Orbiter (TGO), a joint mission of ESA and the Russian space agency Roscosmos, was inserted into orbit around Mars in 2016; and the Mars Orbiter Mission (MOM, also called Mangalyaan), launched by the Indian Space Research Organisation, arrived at Mars in 2014. ESA also sent the British-built Beagle 2 Lander to Mars as part of the Mars Express mission, though it failed before reaching the surface. The ExoMars Lander, Schiaparelli, also failed to land safely in 2016.

bar for publication in *Science* is very high, as high as in any journal—
only about 10 percent of manuscripts submitted to *Science* survive
this process and are published. The McKay et al. paper rose above that
bar. While publication in *Science* does not guarantee that the published results are correct, it does guarantee that a handful of impartial
experts examined the scientific claims rigorously, and it does give the
published results a great deal of credibility (and visibility).

In practice, *Science* likes to publish papers that will generate a great
deal of interest beyond the narrow community of scientific experts
in the niche field associated with the research project. High visibility usually means cutting-edge science, and quite often science done
at the cutting edge turns out to be wrong. Papers published in *Science* naturally lead to press releases, newspaper headlines, television
reports, and other forms of publicity. Of course, fossil evidence of
possible ancient life on Mars, being almost unimaginably big news,
spawned a press conference and quickly led to worldwide fame and
notoriety for the authors.

Virtually all scientists who studied ALH 84001 over the decade
that followed the McKay-team press conference would do so with
open minds as to whether the McKay team's claims for ancient biogenic activity on Mars, as seen in the evidence contained in a Martian
meteorite, might be correct. Within the broader scientific community, the McKay team's claims were immediately met with a strong
and healthy skepticism. Battle lines were quickly drawn and scientists chose sides.

The meteorite community had been down the extraterrestrial-life-
in-a-meteorite road before, three decades earlier, with the Orgueil
meteorite. This previous episode actually began in the nineteenth
century, when a meteorite shower occurred over Orgueil, France, on
May 14, 1864. Only seventeen days later, a claim was put forward
that an analysis of that meteorite revealed it to contain some kind of
chemical residue that resembled humic acid, which is derived from
the decay of organic matter. The author of this claim suggested that
the presence of humic acid in the Orgueil meteorite implied that the
parent body of this meteorite must have had living things in or on it.

According to Edward Anders, then of the Enrico Fermi Institute at the University of Chicago, who retells this story, the Orgueil

meteorite fell just a month after Louis Pasteur's famous lecture of April 7, 1864, entitled "On Spontaneous Generation," which Pasteur presented at the Sorbonne in Paris. Pasteur thoroughly debunked the idea that spontaneous generation occurs. "No," Pasteur thundered, "there is not a single known circumstance in which microscopic beings may be asserted to have entered the world without germs, without parents resembling them. Those who think otherwise have been deluded by their poorly conducted experiments, full of errors they neither knew how to perceive, nor how to avoid."[10] Anders suggested that Pasteur's report "may conceivably have inspired a person of the proper disposition into playing a little practical joke on the scientists." Perhaps the practical joker was also less than thrilled with Charles Darwin's evolutionary arguments, presented only a few years earlier in Darwin's *Origin of Species*, and was attempting to stir up anti-Darwinian sentiments. "Somehow," Anders concluded, "the plot failed, and the contaminated stone went unrecognized for 98 years."

News of organic matter in the Orgueil meteorite became known widely enough to have inspired Swedish playwright August Strindberg, who while living in France in 1887 wrote *The Father*. The father, a retired captain and active scientist, claims to have "submitted meteoric stones to spectrum analysis, with the result that I have found carbon, that is to say, a clear trace of organic life." His wife uses the captain's apparently crazy idea as part of a scheme to have a doctor declare him insane. After the doctor arranges for the captain to be placed in a straitjacket in preparation for having him removed to an asylum, the captain suffers a stroke and dies.[11]

After almost a century, the practical joke reemerged in 1962 when Bart Nagy, a chemist at Fordham University, and his collaborators began their study of the Orgueil meteorite. They found evidence of extraterrestrial life, or so they thought, in one fragment of Orgueil, in the form of "fossilized, organic, organized structures, that are not likely to be minerals, organic artefacts or terrestrial, microbiological contaminations." They went further, writing in their paper in *Nature*, "At present, we are of the opinion that the organized elements are microfossils apparently indigenous to the meteorite parent body."[12] For two years, a scientific battle raged until Anders and his colleagues ended the war of ideas with a tour de force paper in *Science* whose title says everything one needs to know: "Contaminated Meteorite."[13]

"There can be little question," Anders wrote, "that stone No. 9419 has been contaminated. Most probably, the contamination occurred in 1864, shortly before or after the meteorite was put in the museum." Somebody added coal fragments and some local plant material (identified by Anders as identical to the perennial reed plant *juncus conglomeratus*, which grows abundantly throughout Europe and is commonly known as Compact Rush) to the meteorite, moistened it, and glued the contaminants to the specimen. Since coal was not used for household heating in France in the 1860s and likely must have been obtained from a blacksmith's forge, the conclusion that the contamination was intentional is highly plausible. Anders strongly implied that the contamination was done as an intentional hoax, perpetrated in 1864 as a practical joke on the French scientific community.

The Orgueil hoax failed miserably in the sense that it did not trap its intended audience of mid-nineteenth-century French intellectuals and went undiscovered for a century. This hoax also succeeded remarkably well in demonstrating the robustness of the scientific process. Scientific results must be reproducible by impartial members of the scientific community. An especially important and remarkable discovery will be tested, including by some whose scientific agendas are simply to prove that the original discoverers are wrong. By challenging an idea that looms large in both the international and public domains, challengers can establish their own international reputations by debunking claims made by others.

If ALH 84001 contains evidence for ancient biogenic activity on Mars, that evidence must survive extraordinary scientific scrutiny, as such an extraordinary claim should have to do. Beginning as early as the first press conference on ALH 84001 in 1996, the evidence presented by McKay and his colleagues in their paper in *Science* was challenged. Sometimes those challenges were offered respectfully and through the scientific process via papers and debates at conferences, but other times publicly via mudslinging and name-calling. The stakes were high; after all, the opportunity to be the first humans to discover fossil evidence of life on another planet occurs only once.

controversy

What, then, is the evidence for ancient life on Mars, in addition to the indisputable fact that the meteorite itself came from Mars? McKay and his colleagues identified four distinct and independent lines of evidence as proof that ancient Martian life-forms had affected the contents of ALH 84001.

First, they saw small (20–40 nanometers* wide), rod-shaped structures composed of carbon-bearing molecules that resemble rod-shaped bacteria. Without any doubt, ALH 84001 contains tubular, rope-like structures that *look like* certain kinds of terrestrial bacteria. But are they fossil bacteria? They are smaller than any bacteria that, in 1996, were known to exist on Earth. Could bacteria of such small sizes exist? The shapes alone do not make the objects fossil bacteria, but the small sizes alone do not mean they cannot be.

Second, they found orange-colored, pancake-shaped, carbonate globules; that is, the globules are rich in minerals that contain the carbonate ion CO_3^{-2}. Intimately associated with these globules, they identified microscopic mineral grains that they believed appeared to be of bacterial origin. That is, on Earth these mineral grains are all commonly manufactured as products of biological activity. Contamination by terrestrial carbonate sources is not considered an issue by any critics, in large part because the carbonate globules, which comprise about 1 percent of the mass of this meteorite, are physically associated with the tubular structures that might be fossil bacteria. Another particularly fascinating aspect of the carbonate blobs is that they almost certainly formed in a liquid water environment, so they formed on a location on Mars that was warmer and wetter than any part of the Martian surface is today.

Third, they found organic (carbon-bearing) compounds known as polycyclic aromatic hydrocarbons (PAHs), which they asserted were formed from the "diagenesis of microorganisms," which is the process by which organic matter turns into sediment. A PAH is a ring-shaped molecule that contains both hydrogen and carbon atoms.

---

*1 nanometer is a billionth of a meter (or 4 hundred-millionths of an inch). A DNA molecule is 2–12 nanometers wide. A human hair is about 50,000–100,000 nanometers in diameter.

More than one hundred different PAHs have been identified on Earth, and they can be of either biological or nonbiological origin. Some are manufactured; others form from the incomplete combustion of organic matter (e.g., cooking meat or burning tobacco, coal, or oil) or from the slow but natural decomposition of dead organisms. Notably, PAHs "are not produced by living organisms and do not possess any specific role in living processes."[14] But at the August 7 press conference, Richard Zare, who led the laboratory team that found and studied the PAHs, stated definitively, "It [these PAHs] very much resembles what you'd expect when you have simple organic matter decay."[15]

Finally, they found magnetite crystals coexisting with iron sulfide grains in the carbonate globules. According to the McKay team, these magnetite particles "are similar (chemically, structurally, and morphologically) to terrestrial magnetite particles known as magnetofossils. . . . Some of the magnetite crystals in the ALH 84001 carbonates resemble extracellular precipitated superparamagnetic particles produced by the growth of anaerobic bacterium strain GS-15."[16] That is, they look like crystals that are made by a terrestrial species of bacteria. Of great importance to the case for the biological origin of the magnetite particles is the McKay team's assertion that the magnetite crystals and the iron sulfide grains should not occur together naturally; they can, however, be found together in some living things on Earth, where biogenic processes force the creation and preservation of both.

How have these four lines of evidence of life in a Martian meteorite stood up under the test of time and under the withering scrutiny of skeptics? Not well.

Are the tubular structures that look like bacteria actually bacteria? They are small, very small, almost certainly too small, according to the very strong consensus opinion of experts. In 1998, the National Research Council convened a National Academy of Sciences panel to determine, to the best understanding of modern science, the limiting size of very small microorganisms.[17] "Free-living organisms," the panel concluded, "require a minimum of 250 to 450 proteins along with the genes and ribosomes necessary for their synthesis. A sphere capable of holding this minimal molecular complement would be

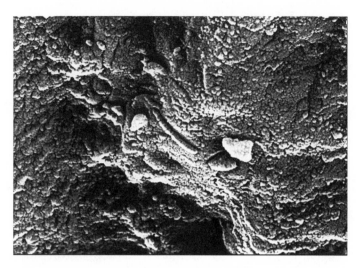

Figure 11.3. High-resolution image, obtained with a scanning electron microscope, of the inside of the meteorite ALH 84001. At the center of this image is a tube-like structure whose diameter is less than 1/100th that of a human hair and that is located in a carbonate globule inside the meteorite. The scientific team that performed the original investigation of this sample argued that the tube-like structure was a fossil of a bacteria-like life-form. Image courtesy of NASA's Johnson Space Center/Stanford University.

250 to 300 nm in diameter, including its bounding membrane." As for the smallest bacteria yet observed, "bacteria with a diameter of 300 to 500 nm are common . . . smaller cells are not." The National Academy panel, however, did not completely preclude the possibility that primitive microbes might have once been smaller, perhaps as small as 50 nanometers in diameter. Even that size, however, is larger than the diameters of the worm-shaped structures in ALH 84001.

The consensus of the experts appointed by the National Academy was challenged by a handful of scientists who thought they had a better and different answer. Their challenge was rooted in a discovery made in the hot springs at Viterbo, Italy, in 1989 by Robert L. Folk, a geologist at the University of Texas at Austin, who found therein what he claimed were nanobacteria. The supposed Viterbo hot springs nanobacteria are exceedingly tiny, ranging in size from 10 to 200 nanometers. With McKay's announcement of his discovery of possible 20–40-nanometer fossils in a Martian meteorite, Folk's nanobacteria suddenly became tangible proof of the reality of such small living things on Earth. Alas, the overwhelming abundance of evidence that had emerged by 2010 reveals that the materials uncovered by Folk have been "conclusively shown to be nonliving nanoparticles crystallized from common minerals and other materials in their surroundings."[18]

Additional arguments have been put forward regarding the speciousness of identifying a structure as biological simply because of its shape. Many studies have shown that "morphology alone is a poor and ambiguous indicator of biogenicity."[19] Many common minerals can resemble biological structures, like those structures seen in ALH 84001. Some of these are even created as artifacts in the process of preparing materials for examination under certain microscopes, as the materials being prepared for study must be painted with special coatings that allow them to properly respond to the examination technique.

What about the PAHs? PAHs are not special or unique. They exist throughout the length and breadth of the universe. Astronomers have identified them in interstellar clouds, in the atmospheres of red giant stars, and in the expanding shells of dying stars (planetary nebulae). Closer to home, meteoriticists and astronomers have found them in meteorites known as carbonaceous chondrites, on the surfaces of asteroids, and in the atmosphere of Saturn's largest moon, Titan. In fact, not finding PAHs might be more difficult than finding them.

The PAHs in ALH 84001 have been studied intensively, and those studies have yielded a variety of explanations and challenges. One fundamental challenge is whether the PAHs are Martian in origin or are contaminants, either terrestrial or extraterrestrial. According to the McKay team, the PAHs in ALH 84001 were deposited when a low-temperature fluid penetrated the cracks in the rock. But not everyone agrees. Jan Martel, of the Laboratory of Nanomaterials at Chang Gung University in Taiwan, wrote in a major summary review published in the journal *Annual Review of Earth and Planetary Science*, in 2012, that we know some other meteorites "harbor PAHs similar to the ones found in ALH 84001," and some studies "have concluded that the vast majority of the organic molecules found in this meteorite do, in fact, represent terrestrial contamination." Whether any of the organic material in ALH 84001 originated on Mars, he wrote, "is still up for debate."[20]

As for the carbonate globules that are associated with the PAHs, they are problematic as well. If the carbonates are of Martian origin, as McKay and his colleagues proposed, then a fluid must have gently washed organic material into the rock, whence the globules subsequently precipitated out of solution into the cracks within ALH 84001. In this case, both the temperature of the fluid that penetrated the rock and the environment in which this event occurred matter. One early challenge argued that the carbonates formed in a CO_2-rich fluid environment at very high temperatures (>650°C, or >1200°F), as a consequence of an asteroidal impact with the Martian surface.[21] In a 1998 study, Laurie Leshin, of UCLA, offered two other high-temperature (above the boiling point of water) formation alternatives—formation at 125°C (257°F) in a water-rich environment or formation at a temperature above 500°C (900°F) in a CO_2-rich fluid environment. "Neither," she concludes, "is consistent with biological activity."[22] Another study, conducted in 2005 by Edward Scott and his colleagues, of the Hawai'i Institute of Geophysics and Planetology, claimed to demonstrate "that the carbonates in ALH 84001 could not have formed at low temperatures, but instead crystallized from shock-melted material. This conclusion," these authors wrote, "weakens significantly the arguments that these carbonates could host the fossilized remnants of biogenic activity."[23] Most recently, however, Itay Halevy, of Caltech, determined that the carbonate minerals "precipitated at a temperature of approximately 18°C (64°F) ... pointing to deposition from a gradually evaporating, subsurface water body." Halevy also concluded that "Though the mild temperatures point to an environment that might be considered habitable, the presence of water was also ephemeral, suggesting a time frame probably too short for life to have evolved de novo [edit: i.e., starting from non-living material]." While the arguments continue to go back and forth as to the temperature of the environment in which the organic material that eventually became the carbonate globules was deposited, none of the models support the possibility that life was involved, and all of them are consistent with explanations for their formation involving "nonbiological precipitation of minerals from supersaturated aqueous solutions."[24]

In the end, we are left with only the magnetite crystals as evidence that might support the life on Mars hypothesis. Perhaps these crystals provide compelling evidence for Martian life. On Earth, some bacteria build within themselves chains of magnetic crystals that they then use to orient themselves with respect to Earth's magnetic field. Numerous studies by members of McKay's research team yielded reports that the magnetic crystals in ALH 84001 were similar to those found in magnetotactic bacteria on Earth. Many of the arguments they have used to assert the biological origin of these magnetic crystals—that the crystals are extremely similar in their sizes and shapes and crystallography to those found for magnetic crystals in magnetotactic bacteria on Earth—have not, however, withstood the test of multiple scientific teams examining the same crystals. As Martel noted in the summary study in 2012, "Other researchers have shown that there exists considerable structural, morphological, and crystallographic variability in the magnetite crystals found in various species of magnetotactic bacteria, suggesting that it is difficult to confirm a biological origin for magnetite particles simply by comparing them with the magnetite crystals observed in terrestrial bacteria."[25] Other studies have questioned whether the magnetic crystals in ALH 84001 form chains. "Taken together," Martel concluded, "these results call into question the hypothesis that the magnetite crystals found in ALH 84001 are of biogenic origins. . . . it can be safely said that the verdict on the presence of exobiological life on Mars cannot be reached on the basis of this thin line of evidence alone."[26]

In 2003, Allan Treiman, of the Lunar and Planetary Institute, in Houston, Texas, wrote cautiously in a NASA report, "The hypothesis of McKay et al. has not been validated. . . . nearly all the data on ALH 84001 and on Earthlife developed since 1996 is [sic] not consistent with the claims, arguments, and hypothesis of McKay et al."[27] A decade later, Martel wrote similarly, "The main arguments that have been used to support the case for past life in the ALH 84001 meteorite can be best explained alternatively by nonliving chemical processes."[28]

The larger scientific community has reached an equilibrium, if not a consensus, on how to interpret the evidence found in ALH 84001. Those few who believe that the mineralogical evidence in

ALH 84001 demonstrates that life once existed on Mars continue to believe they are right and continue to do research that they believe generates additional support for their position. After all, while they have not provided incontrovertible proof for evidence of past biogenic activity in a Martian meteorite, they argue that their scientific opponents have not provided absolute proof that they are wrong. Meanwhile, every new measurement they make that suggests the magnetite crystals could be of biogenic origin inspires others to dig deeper, to make sure the modeling equations are correct (or to correct them), to fine-tune our understanding of how the geochemical reaction steps proceed (or to correct our understanding of these processes), and to make better measurements with their electron microscopes. With virtually every new positive report on the biogenic origin of the magnetite grains, the naysayers generate their own new scientific information that allows them to volley the conclusion of "yes, the minerals are evidence of life" back across the net with a firm "no, they are not." As this cycle of healthy scientific debate continues, the science gets better and our knowledge of Mars, of Martian geochemistry and of ancient paleontology improves.

The work surrounding ALH 84001 is a case study of highly skilled scientists doing science extremely well, not an example of incompetent scientists doing science badly. While the interpretation by McKay and his team of their data was immediately controversial, neither the data published by the McKay team nor the remarkably high quality of those data were ever in question. Neither are the data that they and others continue to produce.

This case is also an example of how the many influences—politics, media, funding, fame—that scientists would like to think are external to the pursuit of truth that drives scientists can shape how science is reported and even done. Scientists often have no choice but to follow the money that funds their research; meanwhile, NASA directs money to research activities that the public and Congress support. The result is that nonscientific reasons often motivate what science is funded. In this instance, the big publicity splash resulting from a discovery made in a single meteorite found in Antarctica has had an enormous impact on science: big money from NASA has been channeled into meteoritics, into searches for extremophiles, and into the race to Mars, all because of the enormous potential effects of the discovery of life in one meteorite.

In this particular case, over a period of more than a decade the scientific method—test, retest, and then test again—has brought forward a consensus. Most of the original evidence for ancient life on Mars as found within this meteorite has not survived the rigors of scientific challenges. The full weight of the argument now appears to rest on whether the magnetite grains could have been produced by inorganic means or only by biological processes.

We do not yet have the extraordinary evidence we should demand that would lead us to agree that ALH 84001 presents evidence of ancient biogenic activity on Mars. The search goes on.

12

methane and mars

While the evidence for biologically produced methane in the atmosphere of Mars is no less controversial than the evidence that Martian nanobacteria produced magnetite grains and deposited them in a Martian rock that later became a meteorite from Mars, the Martian methane debate has had great staying power. For half a century, astronomers have been searching for evidence of methane in the atmosphere of Mars. Why? Methane is a simple molecule that packs a very powerful astrobiological punch: in the oxygen-rich, hydrogen-poor Martian environment, almost no methane should exist without being actively created by living things. Therefore, the discovery of more methane in the atmosphere of Mars than could exist there without life, as is the case in the atmosphere of Earth, could be unequivocal proof that life exists or existed on Mars.

Methane is a colorless, odorless gas that is the simplest possible molecule composed of both and only hydrogen and carbon atoms. (Because methane is odorless, energy companies add an odorant, mercaptan, which smells like rotten eggs, to the methane used for cooking and heating in homes; the smell of the mercaptan helps in detecting gas leaks.) In the language of chemistry, methane is a hydrocarbon. One methane molecule contains one carbon atom bonded to four hydrogen atoms (CH_4). Because hydrogen is the most abundant element in the universe and because, after helium, carbon is the third most abundant element in the universe, methane also exists everywhere in the universe where it can survive; it cannot, however, survive everywhere.

Methane is a common and familiar gas on Earth, as it is the most abundant type of molecule found in natural gas extracted from oil and gas wells, shale deposits, and coal seams. It is the product of the decomposition of organic matter, mostly from ancient marine microorganisms deposited in terrestrial sediments over the last half billion years, that was then subjected to immense heat and pressure for millions of years under Earth's crust. The natural gas used in home heating and cooking is almost pure methane (the gas used in most common barbecue grills is propane, C_3H_8, which is also extracted from the same underground gas reservoirs). On Earth, virtually all the methane, whether buried in underground deposits or free in the atmosphere or trapped in methane clathrates in permafrost, *is biological in origin*. Methane on Earth is a clear chemical signature that life exists (or existed in the ancient past) on Earth.

Methane is a fragile molecule. Because it cannot survive if the temperature is too hot (above about 1500 K), methane is not present in the atmospheres of stars, though it is present in the atmospheres of many brown dwarfs (brown dwarfs are more massive than planets but less massive than stars). Methane also requires a hydrogen-rich (reducing) environment in order to survive and cannot endure in the presence of free oxygen atoms or in an environment rich in oxygen-bearing molecules, such as oxygen (O_2) or carbon dioxide (CO_2). In such an (oxidizing) environment, the carbon-hydrogen bond in the methane molecule is not strong enough to resist the chemically aggressive nature of oxygen. As a result, oxygen atoms rip apart the carbon-hydrogen bonds and then attach themselves separately to both the carbon and the hydrogen atoms. These chemical reactions favor the formation of carbon dioxide (CO_2) and water (H_2O) molecules at the expense of methane. Because the Martian atmosphere is made of 96 percent carbon dioxide, it is a very strongly oxidizing environment. Methane in the Martian atmosphere simply cannot survive for long. Any ancient methane in the atmosphere of Mars would long ago have been converted to carbon dioxide and water; therefore, any current reservoir of Martian atmospheric methane must have been actively produced or released from an underground reservoir in the very recent past.

Methane is at risk in the atmosphere of Mars for another reason. The methane molecule cannot hold itself together when exposed to ultraviolet light. The bonds that attach the atoms to each other in the methane molecule absorb ultraviolet light energy and as a result are destroyed. Interstellar space, for example, which is bathed by ultraviolet photons produced by hot stars, is devoid of methane, except in the densest cores of giant molecular clouds where the outer layers of these clouds can shield small, internal pockets of methane from the ravages of ultraviolet light. The atmospheres of some planets and some of their moons in the outer solar system also are capable of protecting their methane reservoirs and thus can sustain methane-rich environments. The atmospheres of the planets Uranus (2.3 percent methane) and Neptune (1.5 percent methane) have significant amounts of methane because those planets are rich in hydrogen, while Saturn's moon Titan has methane lakes and icebergs, methane-rich mud, and a methane-rich atmosphere. The surfaces of Pluto and Neptune's moon Triton are richly tiled with methane ice, while both Pluto and Triton also have methane in their atmospheres (Pluto even shows evidence for methane frost and snow at high elevations[1]).

The atmosphere of Earth has only a tiny amount of methane (1800 parts per billion, or 0.00018 percent of Earth's atmosphere), and our ozone layer only partially protects what little we have. The small amount of ultraviolet sunlight that penetrates Earth's ozone layer slowly destroys methane atoms in Earth's atmosphere, such that a typical methane atom can survive for only about twelve years. Once sunlight dissociates the carbon atom from the hydrogen atoms, in the oxygen-rich atmosphere of Earth the free carbon atom bonds with one oxygen molecule (O_2) to form carbon dioxide, and the two hydrogen molecules ($2H_2$) react with one oxygen molecule to form two molecules of water.

As for the origin of the methane in Earth's atmosphere, a tiny amount is produced through natural geological processes, as it is belched out of volcanoes or through a process called serpentinization, in which water, heated by the magma at mid-ocean ridges, reacts with iron-rich and magnesium-rich rocks to form the mineral serpentine. The hydrogen atoms liberated from the water molecules in this process react with carbon dioxide dissolved in ocean water to form methane.

Most of Earth's atmospheric methane, however, has a biological origin, from among the following sources:

- the effluence that results from the digestive processes of termites (possibly as much as 15 percent of the total methane content of the atmosphere[2])
- ruminant livestock, such as cattle, buffalo, sheep, goats, and camels (estimates range as high as 20 percent)
- decomposition of organic waste in landfills, wetlands, waste-water treatment facilities, and manure management systems (possibly as much as or greater than 30 percent)
- microbes that feed off organic material in rice paddies (probably more than 6 percent and perhaps as high as 12 percent)
- production, combustion, and distribution of fossil fuels, including coal mining (most likely more than 15 percent and perhaps as high as 30 percent)[3]

As a result of past and present biological activity, Earth's atmospheric methane level, minute as it is, is about one million times greater than the level that volcanic and other geological sources alone can generate. Thus, if life did not exist and had never existed on Earth, the methane level in Earth's atmosphere would be much less than one part per billion.

Unlike Uranus, Neptune, and Titan, Mars is not one of the methane-rich locales in our solar system. Other than a small fraction of a percent of the meteoritic dust that rains down onto the Martian surface, Mars lacks a rich source of methane—unless Mars has life. Martian volcanic activity appears to be extinct; Mars, unlike Earth, lacks a plate tectonic system; also, unlike Earth, Mars has no large, domesticated herds of livestock roaming the surface.

If Mars has a modern, active reservoir of methane, that source remains obscured. And if Mars has a source of methane, we can reasonably ask what will happen to that methane when it passes from rocks or plants or animals into the Martian atmosphere. Though much thinner than Earth's atmosphere, the Martian atmosphere is thick and mobile enough to take any injected methane from any source

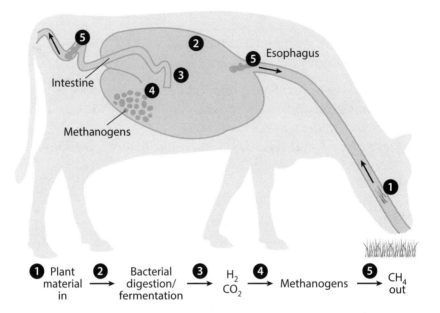

Figure 12.1. Domesticated livestock, especially cows, sheep, goats, and camels, produce methane gas when the bacteria in their digestive tracks helps them digest food.

and spread it globally in only a few weeks' time. Mars also lacks an ozone layer that could shield methane molecules from destructive solar ultraviolet photons, though the abundant CO_2 in the lower Martian atmosphere does act as a weak shield from these destructive photons. Once methane molecules bubble above the bulk of the carbon dioxide in the lower atmosphere, which they will do because they are lighter in weight than carbon dioxide molecules, sunlight will destroy them. The destruction of methane also depends on the presence of oxygen-containing species, especially the OH radical, with which it can react. Because the abundance of OH is so low in the Martian atmosphere, methane can survive for much longer there than it could in Earth's atmosphere. The lifetime for methane in the Martian atmosphere, as a result, is estimated to be 300–600 years, according to the calculations of all Martian atmospheric models.[4]

The Martian atmosphere therefore should contain no methane unless the planet has an active source that is able to continually replenish the atmospheric supply of methane. Furthermore, even

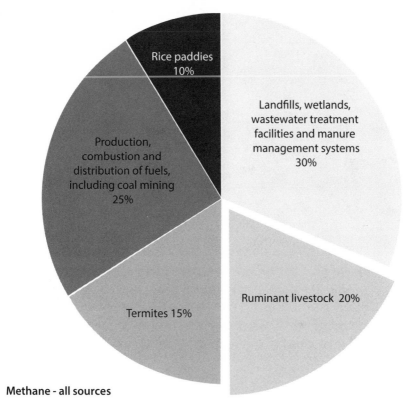

Methane - all sources

Figure 12.2. Pie chart illustrating the sources of methane in the terrestrial environment, all of which ultimately are attributable to past or present life.

if Mars has one or more localized sources of methane, because of the rapid, global circulation of Martian atmospheric gases the entire planet should show about the same amount of methane unless the methane was injected into the atmosphere extremely recently (i.e., within the last few weeks).

Mars does have enormous—but extinct—volcanoes, the largest volcanoes in the entire solar system, in fact. They could inject significant volumes of methane into the atmosphere were they to erupt. But volcanic eruptions do not simply erupt methane. Volcanoes also belch other gases, including sulfur dioxide (SO_2) at a level 100 to 1,000 times greater than that of methane. If Mars were volcanically active, the atmosphere would contain easily detectable levels of

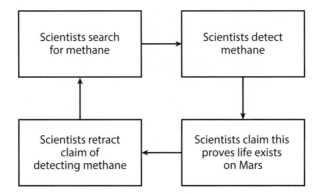

Figure 12.3. The Mars methane science cycle.

sulfur dioxide in addition to methane, but it does not.[5] The absence of sulfur dioxide in the Martian atmosphere provides very strong evidence that the Martian volcanic system is not active at the present time and that any methane, if it exists at all, did not enter the atmosphere through volcanic eruptions.

The story of what we know about Martian methane is simple.

Chapter One: Scientists detect methane.

Chapter Two: Because Mars has no obvious nonbiological source of methane, they or others then claim that the detected methane is evidence for the existence of methane-generating life-forms on Mars. Naturally, all the major newspapers and media outlets report this historic discovery, because it is, in fact, a discovery of unprecedented importance: life discovered on Mars!

Chapter Three: A few months later, they or others retract the claim for the detection of methane. No methane on Mars and no evidence for life on Mars.

Some years later, the Martian methane story begins anew: other scientists detect methane and claim that the methane proves life exists on Mars. More big headlines. Soon, they too are forced to walk back their claims for having detected methane and for having proven life exists on Mars.

More time passes; another possibly specious methane detection hits the front pages. More unsubstantiated claims are made for life on Mars as a new generation of explorers chases a discovery that they hope will reward them with fame (and perhaps fortune). A new generation of reporters, ignorant of the past and eager to report on the Next Big Thing, happily reports on the Next Big Thing.

Maybe one of these reported discoveries was or will be correct. That would be incredible news. The lesson of history, however, is that we may be fooling ourselves.

martian methane discovered?

On October 17, 1966, Dr. Lewis Kaplan, of Caltech's Jet Propulsion Laboratory, stepped to the podium in the Jack Tar Hotel in San Francisco and announced news of a stunning scientific discovery. There, before his colleagues at a meeting of the American Chemical Society, he reported that he and his two French colleagues, Pierre and Janine Connes, two astronomers working at the National Center of Scientific Research and the Meudon Observatory, had detected methane gas in the atmosphere of Mars.

The Kaplan/Connes research team had used an instrument designed by Pierre Connes known as a Michelson interferometer that was capable of seeing details in the light from Mars ten times more clearly than astronomers had previously been capable of seeing. Effectively, they had a better detector and a better telescope but were employing the same technique as that used by Huggins a century before in his search for Martian water. From September 1964 through June 1965, using their new instrument at the Observatoire de Haute-Provence, France, they had measured the contents of the atmosphere of Mars. In doing so they claimed to have observed the spectral signatures of previously undetected gases and to have surpassed the work of previous observers of Mars because of the superior quality of their instrumentation.

What they had detected, exactly and indisputably, were absorption bands in the near-infrared spectrum of Mars. Kaplan and the Connes believed these absorption features were best explained by the presence in the Martian atmosphere of "gaseous compounds containing hydrogen." Such compounds were likely "methane derivatives and perhaps methane itself."[6]

According to an article published in the *Los Angeles Times* the next day, the Kaplan/Connes team's measurement "suggests the presence of previously unreported methane and methane-like material in the Martian atmosphere."[7] *LA Times* science writer Irving Bengelsdorf wrote, at the time, "These observations, if correct, suggest that there may be biological activity on Mars producing methane." Though these astronomers lacked the evidence to prove their case, in 1966 the consensus among astronomers was that the only possible sources of methane on Mars were biological. As a result, Dr. Kaplan most likely did not try to dissuade Bengelsdorf from connecting the dots from methane to life on Mars. The methane supposedly discovered by Kaplan, Connes, and Connes therefore did not merely imply the possible presence of life to the scientists involved in trying to detect methane on Mars. This detection of methane was, in the minds of these discoverers, tantamount to the discovery of life on Mars. Among other outcomes of this premature discovery announcement by the Kaplan team is that the media took control of the story, and the story they told was simply this: methane on Mars equals life on Mars.

Without any doubt, these scientists were confident that the names of the discoverers of life on Mars would someday be written boldly in our history books, and that those names would be Kaplan, Connes, and Connes. Schiaparelli had found canals (well, not really) and was famous, but he had not discovered life. Lowell had found more canals (but no, not really) and claimed he had evidence for life on Mars, and Lowell was now infamous, but he also had not found life. Kuiper's lichens were merely dust. Sinton's algae turned out to be terrestrial heavy water and windblown Martian sand. But these methane results had been obtained with a state-of-the-art Michelson interferometer. The work of Kaplan, Connes, and Connes was modern and surely would stand the test of time. Soon, they reported, they would observe Mars again with an instrument yet one hundred times more sensitive than the one they had used for their observations in 1965.

At a meeting held just four months later, in February 1967, at the Institute for Space Studies of the National Aeronautics and Space Administration in New York City, Kaplan again reported "that the air of Mars contains methane, or methane-based gases, whose presence is difficult to explain unless they are generated by living organisms." At this meeting, he reported that new observations "with Mars much closer had strengthened his belief that the earlier findings are reliable."[8] Kaplan explained that the chemistry of the Martian air "seems to forbid any extended lifetime for methane or its sister gases." Consequently, "something must be continuously providing new methane. On Earth this is done by bacteria that live in the absence of oxygen—as in decaying material in swamps."

Pierre and Janine Connes and Lewis Kaplan had announced their initial discovery in an unrefereed letter, known as an announcement, published in *Science* magazine in August 1966.[9] Unfortunately for them, their conclusion that they had detected methane on Mars was wrong. After their public announcements at the American Chemical Society meeting in October 1966 and the ISS/NASA meeting in February 1967, they never again mentioned their claim for discovering methane, though they also never retracted their claim in print. Probably they hoped that their ill-advised and incorrect announcement would be forgotten. They continued to work together for a few years, even publishing several detailed studies of carbon dioxide in the Martian atmosphere,[10] but they were very careful to never again mention or even cite their own earlier work on methane. Quietly, undetected, almost like odorless methane itself, this first hint for the presence of methane in the atmosphere of Mars dissipated into thin air. Interest in Martian methane, however, continued.

mariner 7

Mariner 7 was launched from Cape Canaveral in Florida on March 27, 1969. Destination: Mars. In 1969, NASA did not yet have the ability to put a spacecraft into orbit around Mars; instead, Mariner 7 flew past Mars on August 5, just five days after the flyby of its twin, Mariner 6. Notably for the study of Martian methane or any other component of the atmosphere of Mars, whether sent to Mars as a flyby or an orbiter, a spacecraft surveying Mars from just above the Martian atmosphere does not have to also peer through Earth's atmosphere. Observers using this kind of data to look for evidence of

Figure 12.4. Launch of Mariner 7 spacecraft atop an Atlas-Centaur rocket from the Kennedy Space Center on March 7, 1969. Image courtesy of NASA.

Martian methane therefore do not have to find a way to remove from the data the overwhelmingly strong signature of terrestrial methane, as did Kaplan and his colleagues. As a result, Mariner 6 and 7 and all future missions to Mars would be capable of making much more definitive measurements of the contents of the atmosphere of Mars than would be possible via almost any observations that might be made from ground-based telescopes.

Mariner 6 returned seventy-five images of Mars and determined that the composition of the atmosphere was about 98 percent carbon dioxide. Mariner 7 obtained 126 spectacular, but grainy, images of Mars, and some infrared spectroscopic data that generated a great deal of excitement.

Just two weeks after astronauts Neil Armstrong and Michael Collins put the first human footprints on the surface of the Moon, NASA was at the peak of its game. First, men on the Moon; now, close-up pictures and measurements of Mars. The world was eagerly awaiting the latest news from NASA and the Mariner 7 Mars science team, and NASA didn't waste any time before providing Mariner science updates. Taxpayers, via congressional appropriations, had funded this mission and NASA, quite reasonably, wanted to update the public as soon as possible on the Mariner 7 discoveries. Without question, front-page news about the Mariner 7 mission would be good for the future of NASA and planetary science. The scientific team members were also eager for their just rewards and the publicity that would ensue after the announcement of their discoveries, having worked for years to build and test and calibrate the instruments that now were generating the data that were at last streaming back across the tens of millions of miles of space that separate Earth and Mars. The press was at the door, pens and television cameras at the ready, prepared to turn a handful of geeky engineers and shy scientists into international personalities and possibly even into award-winning heroes.

Data collected on August 5 were quickly (but insufficiently) analyzed and interpreted. The media hordes descended on the Jet Propulsion Laboratory for a press conference that was held on August 7. The Mariner 7 science team did not disappoint them. Walter Sullivan, of the *New York Times*, convinced his editors to put his article on Mariner 7 science discoveries on the front page of the August 8 issue, with the headline, "2 Gases Associated with Life Found on Mars Near Polar Cap." Big news indeed! And an even bigger mistake.

Dr. George Pimentel, Professor of Chemistry at the University of California at Berkeley, announced that his team, using the Mariner 7

Infrared Spectrometer (IRS), had detected methane and ammonia in the atmosphere of Mars in a localized region above its south pole. The spectroscopic signature of methane had been detected by the IRS in a "strong" absorption band at a near-infrared wavelength close to 3.3 microns. In a strong band, two things are certain. First, the gas is very efficient in absorbing light at that particular wavelength; second, a great deal of the gas is present and is absorbing large amounts of light. In comparison, the ammonia band the IRS team observed toward Mars at 3.0 microns was also present but weaker than the methane line. As for location on Mars, Pimentel stated, "We are confident that we have detected gaseous methane and gaseous ammonia between approximately 61 degrees and 76 degrees south latitudes on Mars."[11]

Having made that measurement, Pimentel and his colleague, Dr. Kenneth C. Herr, also a chemist at the University of California at Berkeley, continued, remarking that the detection "gives no direct clue whatsoever concerning its origin." That, however, did not stop them from speculating: "I have no clue as to the origin of these gases, but if the readings are true—and I believe they are—we have to face the possibility they could be of biological origin."[12] Impressively, not only were they able to speculate about origins, about which they knew nothing, they could extrapolate to deduce the locality of the Martian life-forms: "The geographic localization of the absorptions suggests that their origin is in this hospitable region" [the edge of the polar cap, between 61 and 76 degrees southern latitude, where they claimed there was a reservoir of water].

The euphoria lasted barely a month. On September 11, Pimentel announced at another news conference that the spectral features previously attributed to methane and ammonia were, in fact, caused by frozen carbon dioxide (dry ice). His team's improved analysis of the IRS spectral data thereby "remove[d] one of the final hopes scientists held of finding life on the planet." In his words, the fact that carbon dioxide can mimic the behavior of methane and ammonia was "just a cruel coincidence."[13]

Drawing lessons from hasty conclusions is almost too easy. First, science should not be done by press conference. Second, this kind of science is very hard to do right, and takes extreme care, patience, dedication, and perseverance. To their credit, Pimentel and his team eventually demonstrated they had all of these qualities.* They got the science right. In 1972, after several years of effort, they reported an *upper limit* for methane in the Martian atmosphere of 3.7 parts per million (for easy comparison with later measurements, 3,700 parts per billion).[14]

In the language of science and statistics, an upper limit is not a detection. Instead, it is a way of reporting the degree of accuracy of measurements made in an experiment. Usually a scientist will report an upper limit that is three times greater than the "noise level" in the experiment (a "three sigma" upper limit), though sometimes an upper limit is reported that is only twice the noise level.

Let's return from Mars for a moment so we may better understand upper limits and noise levels. Imagine that you are traveling on a bus from Nashville, Tennessee, to Denver, Colorado, with your pet Great Dane, Fido. While on the bus, you use an old-fashioned bathroom scale to measure the weight of Fido. Your measurement reveals that Fido weighs 150 pounds. How certain are you that Fido weighs 150 pounds and not 152 or 147 pounds?

In order to be sure, you repeat this measurement over and over again and get a different answer every time, with your answers ranging, seemingly randomly, from 145 pounds to 155 pounds. Your confidence in your ability to ever know Fido's exact weight is shaken. Obviously, weighing a Great Dane on an old scale on a bus while moving at high speed on an interstate highway yields an answer with limited accuracy. The limit in accuracy is due partly to the equipment (is the scale calibrated correctly? does the scale work the same way every time?) and partly to variables that you cannot control (bumps on the highway; Fido squirming; Fido's food and water intake (and deposits); your poor eyesight in reading the line toward which the arrow on the scale points). All of these factors contribute to the "noise" in the data. Nothing you do will allow you to get an exact answer, but the measurements you made allow you to calculate

---

*Pimentel Hall on the UC Berkeley campus is now named in his honor. In addition, on May 15, 2017, the American Chemical Society established this building as a Historic Chemical Landmark because the IRS was created and developed here. (http://www.berkeleyside.com /2017/05/24/mars-work-uc-berkeleys-george-pimentel-recognized-national-historic-landmark/)

both an average value for Fido's weight as well as the uncertainty, i.e., the noise, on that measurement.

If you did this experiment with Fido by making 100 measurements and then calculating your answer, you might have found that Fido's weight was 151 ± 2 pounds (read as: 151 plus or minus 2). This answer means that if you were to put Fido on a scale and measure his weight for the 101st time, the likelihood that your answer would be more than 149 and less than 153 (within "one sigma" of the average) is 68 percent; you also can be confident that when you measure Fido's weight for the 101st time, you have a 95 percent chance that your answer will be greater than 147 pounds and less than 155 pounds (within two sigma); finally, 99.7 percent of the time when you make that 101st measurement, you will get a weight in between 145 and 157 pounds (within three sigma). You still don't know for sure exactly what Fido weighs, but you have an incredibly high degree of certainty that Fido's weight is within a known 12-pound range. In the real world, knowing a most likely value and having a quantitative measure of the accuracy with which we know that value is the best we can do.

On your next interstate bus trip, you travel with and weigh—100 times—your sister's plump pet guinea pig, Sassafras. The noise level is dependably the same, ± 2 pounds (the noise has nothing to do with the weight of Sassafras and everything to do with the measuring equipment and the measuring environment), and you find that Sassafras weighs in (the average value of your measurements) at a robust 2.8 pounds. You know that guinea pigs normally don't weigh this much, though Sassafras might. How confident can you be that Sassafras weighs 2.8 pounds? From the statistics of your measurements, you can be 65 percent certain that her weight is between 0.8 and 4.8 pounds, 95 percent certain that her weight is between 0 and 6.8 pounds (you have 100 percent confidence that her weight is not below zero), and 99.7 percent certain that Sassafras weighs less than 8.8 pounds. If you are honest with yourself and understand your data, you would acknowledge that you have no idea if Sassafras weighs 2.8 pounds, since 2.8 pounds is barely greater than the noise in the data. All you know, with great confidence, is that Sassafras weighs less than 8.8 pounds and probably weighs less than 6.8 pounds.

What if you tried this same experiment with your pet gerbil, Orion? Since a typical gerbil weighs only a small fraction of one pound, if you use the same scale on the same bus on the same highway as you used to measure the weight of Fido and Sassafras, all you'll learn is that Orion doesn't weigh enough to register on the scale; however, because of the bounciness of the bus, you still get the same level of noise, ± 2 pounds. You could then, with 99.7 percent confidence, report Orion's weight as "less than 6 pounds." In this case, "6 pounds," which is three times greater than your noise level, would be your "three sigma" upper limit. That's all you know for certain. If Orion weighs 0.2 or 4 or 7 or 12 or 29 ounces, you would get the same answer—less than 6 pounds—when measuring Orion's weight with this scale.

Pimentel and the Mariner 7 team eventually concluded that all they knew about methane in the atmosphere of Mars is that they had 95 percent confidence (their upper limit of 3.7 parts per million was twice their noise level) that the abundance of methane was very small. With 95 percent confidence, out of every million molecules in the atmosphere, fewer than four might be methane, and with 99.7 percent confidence, fewer than six might be methane. Keep in mind that "fewer than 6" could be 5 or 2 or 1 or even 0. Mars might have a very little bit of methane or absolutely no methane at all. Despite the initial publicity fuss, in their robust final analysis the Mariner 7 IRS experiment did not detect methane.

In August 1969, at the first Mariner 7 press conference, Pimentel perhaps should have had the wisdom to step more cautiously. After all, he was one of the team of Berkeley chemists who, in 1965, had shown that heavy water in Earth's atmosphere, not algae on Mars, was likely responsible for the Sinton bands. However, the oppressive historical burden shouldered by anyone studying Mars tended to steer investigators very easily toward incautiousness.

The Mariner 7 methane fiasco was an example of science done poorly by premature press conference, for which NASA, the Jet Propulsion Laboratory, the University of California at Berkeley, the involved scientists, and the media all share responsibility. The news of the retraction made page 3 of the *Los Angeles Times* and page 8 of the *Wall Street Journal*,[15] the latter of which noted, quite fairly, that "the mistaken identification last month of methane and ammonia was a perfect example of the hazards of quick conclusions." The *New York Times* did not report on Pimentel's retraction, which must not have fit within the category of "all the news that's fit to print."

mariner 9

Mariner 9 was the second of the last pair of Mars missions launched in NASA's Mariner mission series, and the imaging results of Mariner 9 were revolutionary. While Mariner 8 failed on launch, Mariner 9 was launched successfully on May 30, 1971, and was inserted into orbit around Mars on November 13, 1971, thereby becoming the first spacecraft to orbit another planet. Just achieving that goal was a spectacular accomplishment.

When Mariner 9 reached Mars, the surface of the planet was obscured by a lingering global dust storm. After about a month, however, the view from this orbiting spacecraft to the surface of Mars became clear. The primary mission of Mariner 9 was to image the entire surface of the planet. When these images were transmitted to Earth, Mars metamorphosed from a distant, unknown planet with a handful of large, lunar-like craters into a well-known world, complete with gigantic volcanoes, a canyon—Valles Marineris—grander by far than our own Grand Canyon, ancient riverbeds, outflow channels, and vast crater-strewn regions. Rather than being a bigger version of Earth's Moon, Mars had the biggest volcanoes, the longest and deepest canyons, and the widest, highest-volume, water-carved valleys and channels in the solar system. Suddenly, Mars was a grand and beautiful world, with a rich areological history waiting to be unveiled.

The initial data from Mariner 9's surveillance of the Martian surface yielded enormous surprises that led science team members to speculate fervently about Martian life. Harold Masursky, of the U.S. Geological Survey and leader of the television examination team, speculated intensely about the origin of the ancient river valleys imaged by the spacecraft camera. Clearly, large volumes of water once flowed on the Martian surface. On that conclusion, his guess about

Martian history was reasonable. He then continued, noting that these discoveries "increase the likelihood that future missions to land on the planet might find some signs of life, or at least fossils of past life." On that point, the historical burden of Mars, and perhaps the interests of the press, had led Masursky up to, if not over, the edge. Project team member Rudolph Hanel, of the Goddard Space Flight Center, reported, "We have not found any signs of life on Mars. We can't expect to. But we have not seen anything that would exclude life." The pendulum of hope for finding life on Mars was swinging back in the favor of "yes."[16]

The infrared interferometer (IRIS) on Mariner 9 was used from late 1971 through much of 1972 to collect an extensive library of spectra of the Martian atmosphere. The most important goals for the IRIS science team were to measure the temperature profile of the atmosphere—that is, how the temperature changes from the surface to the top of the atmosphere—and the temperature and pressure at the Martian surface. The design of IRIS also allowed the Mariner 9 team to measure the abundances of minor atmospheric constituents, including some molecules like methane that could have biological implications.

The world had to wait until 1977 to learn about the Mariner 9 measurements of methane. That year, William C. Maguire, of the Laboratory for Planetary Atmospheres at the Goddard Space Flight Center, published a paper in a refereed scientific journal, without an accompanying press conference, in which he presented his analysis of a carefully selected subset of the Mariner 9 IRIS spectra. Maguire analyzed and averaged 1,747 spectra, selected from the total data set to include only spectra taken after the 100th orbit of the spacecraft. This very reasonable culling of the data was made because the global dust storm on Mars that was present when Mariner 9 arrived at Mars had ended by the 100th orbit, and the dust entrained in the atmosphere had mostly settled back to the surface. Since the dust in the atmosphere decreased the appearance in the spectra of weak atmospheric gases, the spectra taken during the first one hundred orbits were of little scientific value.

In his study, Maguire reported that he was unable to detect any methane. His final result was an upper limit (taken by Maguire to mean a value of twice the noise level) of 20 parts per billion (ppb),[17] a result nearly 200 times lower than the two-sigma upper limit of 3,700 parts per billion reported in 1972 as the final answer from the Mariner 7 mission data. This answer means, very simply, that the best we can say, based on the Mariner 9 IRIS measurements, is that we

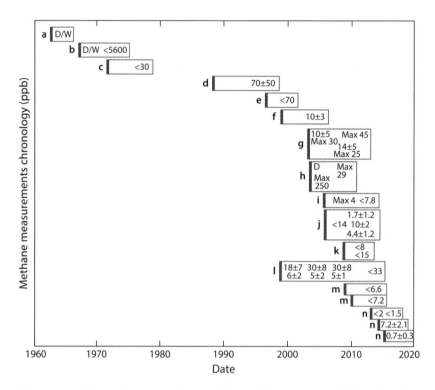

Figure 12.5. A fifty-year chronology of measurements of Martian methane. With one exception ("l"), each box represents one experimental measurement. The left side of each box marks, approximately, the date the measurement was made. The numbers within each box, which indicate the amount of methane reported, are placed approximately at the date(s) publicly reported. When more than one number appears (with the exception of box "l"), that measurement has been reevaluated and the reported answer changed. For example, for "h," the initial report simply noted a detection, a later report gave quantitative values with the range extending up to 250 parts per billion, and a yet later report reduced the upper limit of the range to 29 parts per billion. A detection with no numerical value reported is noted with a "D." If that detection was later withdrawn, retracted, or never published, this final conclusion (or lack thereof) is noted with a "W." All nondetection values (identified with the "<" symbol) are three-sigma upper limits. In box "l," the first set of numbers was for the Martian summer (higher value) and winter (lower value) of 1998–2000, the second set for 2000–2002, the third set for 2002–2004; the final upper limit was a revision that applies to all three of these sets of

are 95 percent certain that the abundance of methane in the Martian atmosphere is less than 20 methane molecules out of every billion molecules and 99.7 percent certain that methane abundance is below 30 molecules per billion. Some methane might exist in the atmosphere of Mars, but not much.

Methane was merely one of several minor atmospheric constituents for which Maguire obtained upper limits, and he offered no commentary on his methane measurements. Methane in the Martian atmosphere, or the absence thereof, was of virtually no interest to astronomers anymore.

numbers. Multiple boxes for the same source ("m" and "n") indicate the same observing team made measurements on different dates that are reported separately in this plot. Data are presented in this figure for observations discussed in later chapters.

a: Kaplan, Connes, and Connes (data 1964–65; reported 1965–66)

b: Mariner 7 (data 1969; published 1972)

c: Mariner 9 (data 1972; published 1977)

d: Krasnopolsky, Kitt Peak National Observatory (data 1988; published 1997)

e: Infrared Space Observatory (data 1997; published 2000)

f: Krasnopolsky, Canada-France-Hawai'i Telescope (data 1999; published 2004)

g: Mars Express (data 2004; published 2004, 2008, 2011)

h: Mumma, 2003 (data 2003; reported 2004, 2006; published 2009)

i: Mumma, Infrared Telescope Facility (data 2006; published 2009, 2013)

j: Krasnopolsky, Infrared Telescope Facility (data 2006; published 2009, 2012)

k: Krasnopolsky, Infrared Telescope Facility (data 2009; published 2011)

l: Mars Global Surveyor (data 1998–2000, 2000–2002, 2002–2004; published 2010, 2015)

m: Villanueva, multiple telescopes (data 2009, 2010; published 2013)

n: Mars Curiosity Rover (data 2013, early 2014, mid-2014; published 2013, 2015)

13

---

digging in the noise

kitt peak national observatory

After a hiatus of a decade, the Martian methane saga emerged from hibernation in 1988 when Vladimir Krasnopolsky, Mike Mumma, and two of their colleagues at the Laboratory for Extraterrestrial Physics at the Goddard Space Flight Center, Gordon L. Bjoraker and Donald E. Jennings, undertook a study of the Martian atmosphere using the 4-meter telescope at Kitt Peak National Observatory in Arizona.[1] Krasnopolsky and Mumma would remain active in the search for Martian methane for three decades, but this would be the first and last time they would work as collaborators rather than competitors in this area of research.

The measurements this team made from a telescope on a mountain on Earth were difficult to make and not as precise as those that could be made from a spacecraft in orbit around Mars. Studying Mars from Martian orbit, however, had not been an option since the Mariner 9 mission ended in 1972 and would not be an option again until Mars Global Surveyor arrived safely in Martian orbit a quarter century thereafter, in September 1997. Though Krasnopolsky and his team obtained these data in 1988, they would not publish the results of their work for nearly a decade, with their paper finally appearing in the *Journal of Geophysical Research* in 1997.

The answer obtained by these observers for the amount of methane in the Martian atmosphere did not simply pop out of their data like a jack-in-the-box. Most importantly, because methane gas in Earth's atmosphere absorbs light at nearly the same wavelengths as the methane in the Martian atmosphere, the terrestrial methane obscures any possible signature of Martian methane in telescopic observations made from Earth. Therefore, Krasnopolsky, Bjoraker, Mumma, and Jennings first had to determine how much methane gas they were looking through as they looked upward through Earth's atmosphere. Their answer: lots, 2,000 ± 100 parts per billion. Next, they had to construct computer models that allowed them to subtract the effects of this huge amount of terrestrial methane from their spectral observations of Mars. Assuming they could accurately remove the signature of terrestrial methane from their spectra, any methane signal that remained in their Martian spectra presumably was due only to methane in the atmosphere of Mars. Ultimately, using measurements of twelve "of the strongest CH_4 lines in the 3.7-micron spectrum," they concluded that the amount of methane in the atmosphere of Mars is 70 ± 50 parts per billion. Note that this amount of methane (70 ppb) is significantly smaller than the *noise level* (100 ppb) in their estimate of the methane in Earth's atmosphere; 70 ppb is also 30 times smaller than the actual amount of terrestrial methane they had to look through, measure, and subtract from their data in order to ascertain the amount of methane in the atmosphere of Mars. They were trying to detect a needle, one camouflaged to look like a stalk of hay, in a haystack. Did they succeed?

Does 70 ± 50 parts per billion mean they detected 70 parts per billion of methane on Mars? No. Just like with the measurement of the weight of the guinea pig Sassafras (2.8 ± 2 pounds), the level of uncertainty (50 ppb) in the measurement by Krasnopolsky and his colleagues is just over 70 percent of the average value of 70 ppb. The uncertainty of "± 50" that they reported tells us that they are 99.7 percent certain that the amount of methane in the Martian atmosphere is less than "70 plus 150" parts per billion; that is, the methane abundance is almost certainly less than 220 parts per billion. We could also assert that we are 95 percent certain that the methane

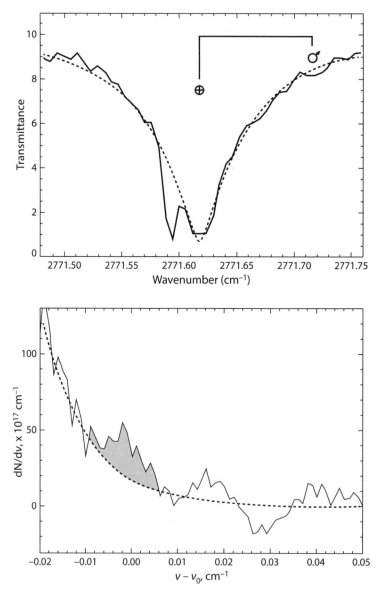

Figure 13.1. Top: Plot of reflected infrared light from Mars ("transmittance") versus wavelength of light (wavenumber). Methane in Earth's atmosphere produces the deep absorption trough centered at 2771.62 inverse centimeters (3.60670 microns; identified with the "Earth" symbol). Because the velocity of Earth is taking it closer to Mars at the time of these observations, the same methane band on Mars is "blue-shifted" to the location marked, at 2771.72 inverse centimeters (3.60657 microns; identified with the "Mars" symbol). Bottom: a close-up of the region centered (at 0.0) on the 2771.72 region from the top figure, with the data effectively

abundance is less than 170 parts per billion. We already knew from the results reported by Maguire from the Mariner 9 IRIS data that the Martian methane abundance was less than 30 parts per billion (with 99.7 percent confidence), so these new results did not meaningfully increase our knowledge about the presence or absence of methane in the atmosphere of Mars.

Krasnopolsky, Bjoraker, Mumma, and Jennings were careful not to claim that they had made a definitive detection of methane on Mars. They wrote that their work "neither confirms nor contradicts the Mariner 9 upper limit." That statement is absolutely right. They continued, however, with a more positive statement: their new results should be understood as "of some interest as an indication of the presence of methane on Mars." That statement falls well short of being absolutely correct. Once again, the historical burden of studying Mars may have led them to overstate their results.

Realistically, an experimenter cannot claim that they can meaningfully distinguish a signal of 70 if 70 is barely 40 percent stronger than the amount of noise in their data (they had a 1.4 sigma result, which no scientist would label as a detection). The Krasnopolsky team results can only be understood "as an indication of the presence of methane on Mars" if one engages in wishful thinking. In fact, it offered no such indication. An upper limit is not a detection or "an indication of the presence" of what the observer is trying to detect. An upper limit is simply an upper limit. End of story. The bottom line is that the attempt to measure the level of methane in the atmosphere of Mars, made using a telescope in Arizona in 1988, yielded only noise. It should never have been reported as anything other than an upper limit (99.7 confidence level) of 150 parts per billion. Future measurements would confirm this more cautious interpretation.

inverted. The dotted line indicates the amount of light that should be measured if (a) there were no methane in the Martian atmosphere and (b) there were no noise in the data. The gray region purportedly is a measure of the amount of methane gas in the Martian atmosphere. This spectral feature is very difficult, if not impossible, to distinguish as meaningfully different from the other deviations, identified as noise. Krasnopolsky et al. argued that it was "of some interest as an indication of the presence of methane on Mars." Image based on Krasnopolsky et al., *Journal of Geophysical Research* 1997.

Nevertheless, the assertion that this discovery was "an indication of the possible presence of methane on Mars" became justification for both Krasnopolsky and Mumma to look again, working separately, using the precious resources of time and equipment on large telescopes in an attempt to confirm their initial results and add more information to the saga of methane on Mars.

the infrared space observatory

Before either Krasnopolsky or Mumma had another chance to search for methane on Mars, their initial claim for the "possible presence of methane on Mars" took a hit. The Infrared Space Observatory (ISO) was a modest-sized (60-cm, or 24-inch, aperture) Earth-orbiting telescope launched in November 1995 by ESA. Its success as a sensitive infrared telescope was largely due to its mirror being cooled by liquid helium to a temperature of just a few degrees above absolute zero (–456°F). In July and August 1997, ISO carried out a spectroscopic study of Mars using the instrument known as the Short-Wavelength Spectrometer, the results of which were reported in 2000 by a team led by Emmanuel Lellouch, of the Observatoire de Paris.[2] The investigators were successful in detecting a number of molecules in the Martian atmosphere, notably carbon dioxide (CO_2), water (H_2O), and carbon monoxide (CO), and studied them in great detail in this experiment.

In searching for methane in the Martian atmosphere, the ISO team had an important advantage over observers using telescopes on the surface of Earth: ISO was above all of Earth's atmosphere, so the light from Mars did not have to penetrate Earth's atmosphere in order to reach the Earth-orbiting ISO telescope. Therefore, unlike Krasnopolsky's team, the ISO team did not have to make the very difficult corrections of their data for the effects of 2,000 parts per billion of terrestrial methane. Their search for methane, conducted at both the wavelengths of 3.3 microns and 7.66 microns, however, yielded no detections, only an "upper limit" of 50 parts per billion, which they note is "comparable [to], but not quite as good as that

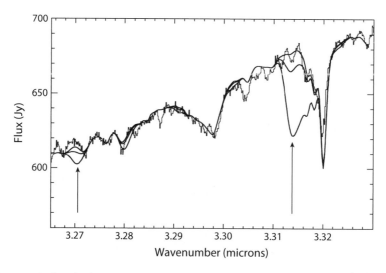

Figure 13.2. Plot of reflected infrared light from Mars (histogram) versus wavelength of light in the region of the absorption band of methane at 3.3 microns, obtained by the Infrared Space Observatory. The three smooth curves drawn through the data are model atmospheres for Mars that assume the presence of no methane (top line), 50 parts per billion of methane (middle line), and 500 parts per billion of methane (bottom line) in the Martian atmosphere. In between 3.29 and 3.31 microns, the three lines are indistinguishable, while at 3.270 and 3.314 microns the three lines show clear differences from each other and from the data. These three model fits show that 500 parts per billion is inconsistent with the data, zero is consistent with the data, and 50 parts per billion is perhaps barely consistent with the data. The other absorption features (these being drops from a steadily rising profile), from left to right, are all due to the Sun, not to Mars. Image based on Lellouch et al., *Planetary and Space Science*, 2000.

of Maguire (1977) from Mariner 9 data (20 ppb)."* The ISO results are also comparable to but slightly better than the Kitt Peak results of the Krasnopolsky team and should also be considered more dependable since the ISO team did not have to correct for the effects of Earth's atmosphere and because these measurements were made in both the 3.3 and 7.66 micron wave bands. The authors of the ISO report note, correctly and with exquisite graciousness, that their results suggest "the tentative detection of CH_4 reported by

---

*This phrasing suggests we should understand the ISO upper limit, like the Mariner 9 upper limit, as equivalent to twice the noise level (95 percent confidence).

Krasnopolsky et al. (1997) from very high-resolution . . . ground-based observations (70 ± 50 ppb) is indeed marginal at best."

canada-france-hawai'i telescope

In 1999, Krasnopolsky partnered with Jean Pierre Maillard, of the Institut d'Astrophysique de Paris, and Tobias Owen, of the Institute for Astronomy at the University of Hawai'i, in a search for Martian methane. With this team, Krasnopolsky had access to a powerful infrared detector system known as the Fourier Transform Spectrometer (FTS) on the 3.6-meter (12-foot) diameter Canada-France-Hawai'i Telescope in Hawai'i. At the high elevation (14,000 feet) of Mauna Kea, astronomers can make much more sensitive infrared observations of the heavens than are possible at Kitt Peak (elevation 7,000 feet), so Krasnopolsky had successfully leveraged his earlier work to gain access to a similar-sized telescope in a much better location for this project than he had a decade before.

He made his new Martian observations in late January 1999, though he would not report his results until 2004. His team announced some preliminary results and their assertion of priority for their discovery of methane in the atmosphere of Mars in an "abstract" released in early January 2004. An abstract is a one-paragraph announcement, presented without any supporting detail, of scientific results that the authors intend to report about in greater detail at an upcoming scientific meeting. Meeting abstracts are submitted weeks to months before dates of the actual meeting and, unlike journal articles published in the professional literature, are not usually subject to peer review. This abstract advertised, very briefly, the scientific results that the Krasnopolsky team planned to present at a meeting of the European Geosciences Union in Nice, France, scheduled for that upcoming April. The title of the talk was firm and clear: "Detection of methane in the Martian atmosphere: evidence for life?"[3] Yes, they were claiming a definitive detection of Martian methane, and they also were attempting to suggest a connection between atmospheric methane and life on Mars. Complete results were presented a few months later in a paper with the same definitive title, submitted to the journal *Icarus* on March 29, 2004, and accepted for publication on July 1, 2004.[4] Because of a long, drawn-out struggle among several research teams for priority for making the first discovery of Martian methane, these dates are important.

Krasnopolsky's observing protocol was difficult but well designed. Because the observations were made from the ground, he had to look through Earth's thick atmosphere, albeit from the summit of Mauna Kea. In the region of the spectrum under study, at least 24,000 different absorption lines of terrestrial methane contaminate the observed spectrum of Mars. Though most of these lines are very weak in Earth's atmosphere, they exist identically in the atmospheres of both Earth and Mars. In order to prove that any detected absorption lines due to methane are of Martian origin, Krasnopolsky had to separate the Martian methane lines from the terrestrial methane lines. The key to this technique was one used a century earlier by W. W. Campbell in looking for water in the Martian atmosphere: Use the Doppler shift.

Krasnopolsky, like Campbell, took advantage of the fact that Earth and Mars orbit the Sun at different velocities. When this difference is large, the Martian methane lines would be Doppler shifted away from the terrestrial methane lines just enough to allow an observer to distinguish between them. That, at least, is the idea.

Planets closer to the Sun experience a stronger gravitational pull from the Sun than planets that are more distant from the Sun. Consequently, planets in small orbits travel through space at higher speeds than planets in larger orbits. Earth, being in an orbit that is only two-thirds the size of Mars's orbit, orbits the Sun at a faster speed (average orbital speed of 18.5 miles per second, or 29.8 kilometers per second) than Mars (average orbital speed of 15.0 miles per second, or 24.1 kilometers per second). As an analogy, think of Earth and Mars as runners on an oval racetrack, with Earth in lane 3 and Mars a bit outside of Earth in the slightly longer lane 4. With Earth running faster than Mars in a shorter lane than Mars, sometimes Earth is behind and catching up to Mars, sometimes Earth is right next to and passing Mars; and sometimes Earth is ahead of and pulling away from Mars. At the time of Krasnopolsky's observations in January 1999, the motion of Earth relative to Mars was 11.05 miles

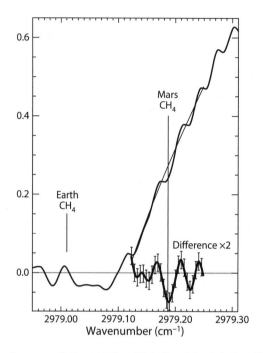

Figure 13.3. Plot of amount of reflected infrared light from Mars (wiggly, solid line that begins at the left, near 0.0, and curves toward the top right corner) versus wavelength of light (wavenumber). The Martian methane line is slightly blue-shifted (toward a higher wavenumber) from the central location of the terrestrial methane line because of the relative velocity of Earth toward Mars. Bold line with associated error bars in bottom right corner is a close-up of the data plotted on the reflected light profile, at the same wavenumbers. The data clearly show nothing but noise in most of the bumps between 2979.10 and 2979.25 inverse centimeters. According to Krasnopolsky, however, the apparent drop in the signal centered at 2979.19 inverse centimeters (3.35662 microns) is real and is due to Martian methane. Image based on Krasnopolsky et al., *Icarus*, 2004.

per second (17.79 kilometers per second), and Earth was catching up to Mars so that the distance between Earth and Mars was getting smaller. Consequently, any Martian methane lines would be blue-shifted by 11.05 miles per second relative to any terrestrial methane lines. In principle, this Doppler shift might be large enough to enable Krasnopolsky to distinguish spectral lines due to Martian methane from spectral lines due to terrestrial methane.

Krasnopolsky, Maillard, and Owen claimed to have detected the supposed methane lines in the Martian spectrum by first finding the spectral lines due to methane in Earth's atmosphere. They then looked for signatures of Martian methane at distances in the spectrum equivalent to a shift from the center of the terrestrial methane lines of –11.05 miles per second. These positions would show up on what spectroscopists would call the "blue wings" of the terrestrial methane lines. There, on the blue wings, they found twenty very tiny bumps that were all in about the right places to be Doppler-shifted Martian methane lines. These bumps were all very small, on the level of noise in the data; however, because their locations were close to where one might expect signals for Martian methane, they argued they were signal rather than noise. After carefully using computer simulations to "fit" methane emission lines for different amounts of methane in the atmosphere of Mars to these bumps, the team argued that the signal (implied by the existence of these bumps) exceeded the noise by just enough to make the case for having detected methane. Krasnopolsky's new answer: the Martian atmosphere contains 10 ± 3 parts per billion of methane.

If this measurement is correct (and as a three-sigma detection, though marginal it plausibly is), and both the title and conclusions of the paper assert that this is a firm detection of methane in the atmosphere of Mars, it is a spectacular and important discovery. Remember what these signal and noise numbers mean: we know that, were Krasnopolsky to repeat his experiment, he would have a 68 percent chance of finding the methane abundance as low as 7 and as high as 13 parts per billion and a 99.7 percent chance of finding that the methane abundance could be as low as 1 or as high as 19 parts per billion.

The newly measured result of 10 ± 3 parts per billion is consistent with the final Mariner 7 measurement (99.7 percent confidence of less than about 5,600 parts per billion), the Mariner 9 measurement (99.7 percent confidence of less than 30 parts per billion), and the ISO measurement (99.7 percent confidence of less than 75 parts per billion). The new results are also consistent with the 1988 measurements (reported in 1997) made by Krasnopolsky and his collaborators at Kitt Peak National Observatory if those results are understood as a measured upper limit (99.7 percent confidence level) of 220 or even 150 parts per billion rather than "as an indication of the presence of methane on Mars."

Could both the 1988 and 1999 measurements be right, assuming the 1988 measurement was, in fact, a detection rather than an upper limit? Could the amount of methane in the atmosphere of Mars have dropped from 70 parts per billion to 10 (or even 1) parts per billion over that decade? An imaginative "yes" explanation is possible, but a realistic "yes" explanation is not. In his 2004 paper, Krasnopolsky referred to his own measurements from 1988 as a "weak signal of possible methane" but otherwise makes no attempt to explain the discrepancy between the 1988 measurement and the 1999 "detection." Clearly he was distancing himself from the earlier results and his own over-zealous interpretation of those results.

As calculated by Krasnopolsky, Maillard, and Owen in a result that has become widely accepted as correct, destruction of methane by sunlight in the Martian atmosphere limits the lifetime of methane in the Martian atmosphere to between 250 and 430 years. The short lifetime of methane in the Martian atmosphere combined with the existence of 10 parts per billion of methane in 1999 strongly implies that some source has been actively pouring methane into the atmosphere of Mars over the last few centuries; otherwise, the amount of atmospheric methane should be below the limits of detectability. Krasnopolsky, Maillard, and Owen worked through arguments for the volcanic production of methane and for the deposition of methane into the Martian atmosphere from meteoritic dust and comets. They concluded that no abiogenic source of methane can generate the 10 parts per billion of methane detected in their measurements. "Therefore," they wrote, "we have not found any significant abiogenic source of methane and methanogenesis by living subterranean organisms is a plausible explanation for this discovery." If this measurement holds up against the test of time, and if the interpretation that "methanogenesis by living subterranean organisms" is correct, it will prove to be a discovery of almost unrivaled scientific significance.

mars express

The Mars Express mission was the first successful interplanetary mission attempted by ESA. Launched in June 2003, Mars Express eased into orbit around Mars on Christmas Day. The lander, Beagle 2, which separated from the orbiter six days earlier, failed to land safely on the surface of Mars, but the orbiter successfully achieved Martian orbit. One of the instruments on board the orbiter, the Planetary Fourier Spectrometer (PFS), built by Vittorio Formisano of the Istituto di Fisica dello Spazio Interplanetario in Rome, included in its design the ability to search for methane.

Using data from 2,931 different observations covering sixteen orbits of Mars during the months of January and February 2004, the PFS team, led by Formisano and including Sushil Atreya, Thérèse Encrenaz, Nikolai Ignatiev, and Marco Giuranna, announced that they had detected methane in the atmosphere of Mars. Although they did not publish their results in the refereed literature until the end of that year, ESA did put out a press release on March 30, 2004, the day after Krasnopolsky's team submitted their results for their 1999 Mars observations from Hawai'i to the journal *Icarus*. The race was now on to stake a claim for being the first to detect methane on Mars.

The ESA press release headline pulled no punches: "Mars Express confirms methane in the Martian atmosphere." The text of the press release noted that the measurements "confirm so far that the amount of methane is very small—about 10 parts in a thousand million [10 parts per billion], so its production process is probably small."[5] The selection and repeated use of the word "confirms" in the ESA press release was an interesting choice. Exactly what previous measurement were they confirming?

The Formisano team reported their final Mars Express results, covering observations from January through May 2004, in a paper published in *Science* in December 2004. They reported that their data revealed a global average for methane in the Martian atmosphere of 10 ± 5 parts per billion (a 68 percent chance that if the experiment were repeated, they would obtain an answer in the range from 5 to 15, and a 95 percent chance they would obtain an answer in the range from 0 to 20),[6] a result almost identical to that reported by Krasnopolsky (10 ± 3 parts per billion) a few months earlier. They also reported a previously undetected phenomenon associated with Martian methane: as measured by Mars Express, *the amount of methane in the Martian atmosphere changes both with time and location* on

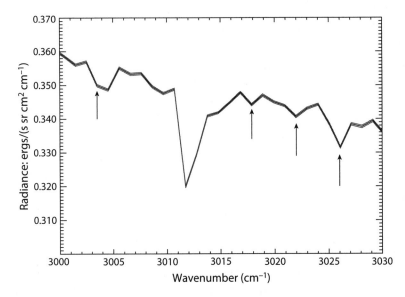

Figure 13.4. Infrared spectrum of Mars obtained by the Planetary Fourier Spectrometer on the Mars Express spacecraft in January–February 2004. The apparent detection of methane in the atmosphere of Mars shows up in the small drop in the spectrum at 3018 inverse centimeters (i.e., a wavelength of 3.313 microns). Three strong, known water vapor lines appear in absorption at 3003.5, 3022, and 3026 inverse centimeters. The deep line at 3012 inverse centimeters is due to the Sun. The middle line is the data. The upper and lower lines indicate the noise level in the data. Image based on Formisano et al., *Science*, 2004.

the planet. The methane levels they measured reached as high as 30 parts per billion and as low as 0 parts per billion. They suggested that these variations from 0 to 30 might be due to differences in the methane abundance with location or with time. Whether the differences are due to spatial or temporal effects, the total time span covered by these observations was only five months, which is short compared to the Martian year (687 days) and short even compared to the length of time for the seasons on Mars to change.

Their detection of methane is based on having identified a single dip in the infrared spectrum of Mars (at the wavelength of 3.313 microns). This tiny dip is identified as methane when a comparison of the observed spectrum is made with a set of computer-generated spectra that show a number of other dips in this same spectral region, most of them due to water vapor. The robustness of the identification of this dip as real and due to methane depends on the accuracy of their computer models, for which they assumed a specified amount of water vapor and dust in the Martian atmosphere. Because molecules and atoms produce multiple spectral absorption lines, not just a single line, detecting methane at more than a single wavelength would generate significantly more confidence in the result, but the PFS instrument could not offer that additional piece of information. One either does or does not believe these results.

The Mars Express results (10 ± 5 parts per billion), would appear to be consistent with the results from Krasnopolsky's 1999 data (10 ± 3 parts per billion), though not with his 1988 measurements (70 ± 50 parts per billion) unless either the amount of methane in the atmosphere of Mars changes dramatically (which the Mars Express team argued does happen) or we understand the 1988 measurements as nothing but noise. According to the calculations presented by the Mars Express team, a comet as a source for the observed methane (e.g., a comet crashing into Mars or Mars sweeping up an enormous amount of material shed by a comet) is an extremely unlikely but not entirely implausible source of the detected methane. Similarly, they suggest that nonbiogenic subsurface sources of methane (e.g., the conversion of CO_2 to CH_4 in a sub-permafrost aquifer) are possible, though not probable. As they carefully stated their case, "We want to stress that the detection of methane does not imply the presence of life on Mars, now or in the past. It is one possibility, but . . . other sources are at least plausible." But the saga of the Mars Express data did not end in 2004.

Further analysis of two full Martian years of Mars Express data, reported in 2008 by A. Geminale, V. Formisano, and M. Giuranna,[7] identified seasonal variations in the methane abundance. They reported values as high as 21 parts per billion in the northern spring-summer and as low as 5 parts per billion at the end of southern summer, with an annual average of 14 ± 5 parts per billion.

They also reported longitudinal variations from 6.5 to 24.5 parts per billion and found that the amount of methane gas is correlated with the amount of water vapor, indicating that the two gases change from solid to vapor together from the same source. "The methane mixing ratio" (that is, the ratio of methane to water molecules in the atmosphere), they wrote, "seems to follow the water vapour diurnal cycle. The most important point for future understanding is, however, that there are special orbits in which [the] methane mixing ratio has a very high value."

The seasonal changes in the amount of atmospheric methane detected in the Mars Express data raise "a very important point: is methane destroyed or is it recycled, being hidden somewhere" for part of each Martian year? "If it is destroyed," they concluded, "the life time cannot be the (300–600) years computed for atmospheric UV photodestruction."

14

here today, gone tomorrow

One more team, this one led by Mumma, would enter the "we detected methane in 2004" competition. Though the Mumma-led team would be the last to actually publish their results in a refereed scientific paper, doing so only in 2009 (after Krasnopolsky, Maillard, and Owen in mid-2004 and the Mars Express team in late 2004), they were chronologically the first to publicly stake a claim to having discovered methane in the atmosphere of Mars, when they did so in 2003. (Mumma's announcement in 2003 may have spurred Krasnopolsky into completing his analysis and writing up for publication the results of his 1999 observing project.)

The May 2003 abstract served as a placeholder for a presentation Mumma would give at an American Astronomical Society Division of Planetary Sciences meeting, which was held in Monterey, California in September of that year.[1] In the first sentence of their abstract, Mumma and his co-authors Robert E. Novak of Iona College, Michael A. DiSanti of NASA's Goddard Space Flight Center, and Boncho P. Bonev of the University of Toledo, made one thing very clear: the "possible" detection reported by Krasnopolsky, Bjoraker, Mumma, and Jennings in 1997 was no such thing. "CH_4 and its oxidation products on Mars have received both observational and theoretical attention but have not been firmly detected," they wrote. Very clearly, as of the summer of 2003, Mumma, a co-author of the 70 ± 50 parts per billion "indicating the possible presence of methane on

Mars" statement, had effectively withdrawn that conclusion and was now in agreement with the ISO team's assessment of that result. The crown of glory that would be awarded to those who would make that first big discovery remained, he was effectively saying, unclaimed.

The rest of Mumma's abstract said absolutely nothing about his own new results, noting only that he had been conducting a "deep search for methane on Mars," and that "details will be presented" at the September meeting. This announcement, however, made clear that Mumma was on the prowl for methane on Mars.

Mumma's team put out a second announcement in late 2004 touting the results Mumma planned to present at the next Division of Planetary Sciences meeting to be held in Louisville, Kentucky that November.[2] This abstract, again presented with colleagues Novak, DiSanti, and Bonev, and also including Neil Dello Russo, of Catholic University, provided some actual details of what he and his collaborators were doing. It also included a no-holds-barred claim—the first three words of the abstract were "We detected methane." The abstract made special note that these observations were made as early as January 2003. Based on the dates of when the detections were made, and having previously dismissed the earlier "possible detection" that he had made with Krasnopolsky, Mumma was now claiming the first detection for himself and his new scientific partners. Of course, earlier that year Krasnopolsky had tried to claim the "first detection," based on his 1999 data, and the Mars Express team had also tried to grab a share of this claim to history, using early 2004 data.

Mumma reported that his team had attempted to detect methane on Mars beginning in early 2003 using three different telescopes: NASA's 3-meter (10-foot) Infrared Telescope Facility (the IRTF), on Mauna Kea, in January and March 2003 and January 2004; the 8-meter (26-foot) Gemini South telescope in Chile, in May and December 2003; and the 10-meter (33-foot) Keck-2 telescope, also on Mauna Kea, some time in 2003. At the 2004 meeting, he announced that his team had detected methane on Mars. Years later, he would present the details of his measurements, reporting that the level of methane in the Martian atmosphere that they had detected was about 10 parts per billion, averaged across the full atmosphere of Mars.

All of the early 2004 announcements about methane on Mars received immediate attention in the popular press. On March 30, 2004, CNN.com reported that "a trio of research teams independently probing the Martian atmosphere for methane have confirmed the presence of the gas."[3] The trio of teams included Krasnopolsky's team, Formisano's Mars Express group, and the Mumma-led partnership. CNN.com quoted Krasnopolsky, who said, "I would say that they [Mars Express] confirm our results." Krasnopolsky expressed his pleasure with the fact that the Mars Express results revealed almost exactly the concentration of methane seen in his own experiment. A spokesperson for the Mars Express team suggested "the first possibility, volcanism, is probably best," while Krasnopolsky told CNN that he favored microbes as the source of the methane. Mumma's team, by virtue of reporting preliminary results in the form of his September 2003 abstract, gained entry into the "trio of research teams," but otherwise was unmentioned.

Notably, all three teams did report similar levels of Martian atmospheric methane. The scientific method, with competing teams using different methods and different telescopes, seemed to be playing out appropriately. Each of three different groups had collected data independently that appeared to confirm and verify the results of the other two groups. An impartial observer, for example the reporter for CNN, should have felt a reasonable degree of confidence that the astronomy community was converging on a reasonable and correct answer: Mars has methane! And Mars might have more methane than could be produced without biological activity!

Although the scientific method seeks consensus, consensus does not always mean correctness. For example, in the seventeenth century, physicist Isaac Newton, astronomers Johannes Kepler and Christian Longomontanus, Reverend John Lightfoot, and Bishop James Ussher all used different methods to deduce that the age of Earth was about 6,000 years.[4] While their answers were consistent both with each other and with expectations of that era, they were all wrong, off by more than four billion years.

Krasnopolsky's research results appeared in a paper published in the journal *Icarus* on August 20, 2004. His paper was the trigger for a *Nature* magazine commentary that appeared on September 21, titled

"Martian methane hints at oases of life."[5] Clearly, the editors of *Nature* thought that the search for and discovery of methane on Mars was a significant scientific accomplishment and that the answer was in hand. The editorial was written by in-house science reporter Mark Peplow, who began his piece by commenting, incorrectly, "researchers have concluded that life is the only plausible source of the [methane] gas. The putative Martians are hiding in a few isolated spots and the rest of the planet is sterile." Indeed, Krasnopolsky had drawn that conclusion, but others were publicly more cautious about the possible source of the possible methane.

Krasnopolsky calculated the volume of methanogenic bacteria currently living on Mars: 20 tons. These bacteria, he asserted, would be "concentrated in just a handful of oases, which would explain why NASA's Viking landers missed any signs of organic chemistry in 1975 and 1976." This analysis might be wrong, but Krasnopolsky had concocted an imaginative hypothesis that would be difficult for anyone to disprove and that continues to animate more recent measurements of methane: Martian bacteria live in only a few isolated locations.

In 2004, Mike Mumma offered his personal note of skepticism about Krasnopolsky's assertion that life on Mars is the source of the methane. "I don't think the community is that impressed [by that idea]," Mumma told Peplow. In Mumma's view, volcanic activity or gas being squeezed out of rocks deep inside the planet were more likely sources. Nevertheless, the *Nature* editorial reported that Mumma had just submitted a proposal to NASA for a space-based infrared telescope that could be launched by 2010, whose purpose included solving the life-on-Mars question once and for all. Whether via gaining access to time on big telescopes or through obtaining a big pot of money to build a very expensive space telescope to pursue answers to his science questions about Mars, the idea that Martian methane might be the critical clue for identifying life on Mars represented an opportunity for Mumma to lead a robotic expedition to Mars, even if he was pooh-poohing Krasnopolsky's nearly identical ideas on the subject. (Mumma's proposed telescope project, however, never got off the ground.)

In a December 2006 interview with Leslie Mullen, published in *AstroBiology* magazine, Mumma explained that he had detected impressively high levels of methane on Mars. "At high latitudes in the north and south, there is much less methane. It's 20 to 60 ppb in the north, and even lower in the south. But it was more than 250 ppb at the equator," he reported. The high level of methane, he found, was confined to the equatorial latitudes from "minus 10 degrees south to 10 degrees north." These results, if upheld, "could define the course of Mars exploration for years to come." In contrast, Krasnopolsky's group and the Mars Express team had both reported detecting much lower levels of methane, about 10 parts per billion. Mumma's results were different, and, he asserted, much more exciting. They also would prove contradictory to the Mars Express results, which would show that the methane abundance is greatest at the north pole rather than at the equator.

No doubt, these claims were bold. Mumma was now claiming to have detected levels of methane in the Martian atmosphere that varied with latitude and that ranged from 20 parts per billion, which is already twice the level seen by the Krasnopolsky and Mars Express teams, up to a phenomenal 250 parts per billion. He also claimed to have found significant amounts of methane at both high northern and high southern latitudes, as well as enormous amounts at equatorial latitudes. Apparently, methane was nearly everywhere on Mars, although the amount varied across the planet.

Mumma's Martian methane results were dramatically different from the methane results reported by the other research groups. Either somebody is right and somebody else is wrong, or the production of methane on Mars is extremely variable in both time and location. As for whether Mumma's new results "could define Mars exploration for years to come," any scientist paying close attention would have to agree. If this much methane exists in the Martian atmosphere, and if the presence of methane is possibly proof of life on Mars, then NASA and ESA are absolutely compelled to reorient everything they are doing in exploring Mars to take account of the fact that Mars is inhabited by living, breathing, microscopic, methanogenic Martians.

In yet another conference abstract, submitted for the Division of Planetary Sciences meeting held in Cambridge, England, in September 2005, Mumma and his team—again including Novak, DiSanti,

Bonev, and Dello Russo, and now also including Tilak Hewagama, Geronimo L. Villanueva, and Michael D. Smith, all of Goddard Space Flight Center—reported that they had identified differences in methane abundances across latitudes ("strong latitudinal gradients"). They suggested that these latitudinal differences in the methane abundances require, even demand, "localized release" of methane and survival of methane in the atmosphere for no more than a few weeks.[6] "I'm shocked by this result," admitted Mumma. "At these two points on Mars, the data imply that there were significant methane releases. . . . This cannot be waved away as a measurement error."[7] Two years later, for the 2007 Division of Planetary Sciences meeting in Orlando, in yet another conference abstract, Mumma—this time with Villanueva, Novak, Hewagama, Bonev, DiSanti, and Smith—reported that the release of methane on Mars favored certain seasons and locations, and that the pattern of releases was episodic.[8]

Finally, in 2009, Mumma published his methane studies in a refereed journal article, which appeared in February in *Science*, with co-authors Villanueva, Novak, Hewagama, Bonev, DiSanti, Smith, and Avi Mandell (of Goddard Space Flight Center). "Before 2003," Mumma began, "all searches for CH_4 were negative."[9] "Since then," he continued, "three groups have reported detections of CH_4." He failed to make clear that prior to 2009 one of those groups, his own, had only talked about their work at conferences and had never submitted their work for peer review in the refereed literature, as had the other two groups. He also credited his own group with five of the nine "publications" noted as reports of previous detections, even though all five of those were unrefereed conference abstracts. This reportage put an interesting spin on the history of methane studies of Mars, as most professional scientists do not count abstracts as publications. They serve a purpose in the scientific process as both announcements and progress reports. But they are not publications in the way that journal articles that have gone through the refereeing and editorial process are publications.

Mumma reported that he had detected variations in the atmospheric abundance of methane both with latitude and with a change in Martian seasons. All of these reported changes occurred in 2003. The mean methane abundance he reported was lowest both north

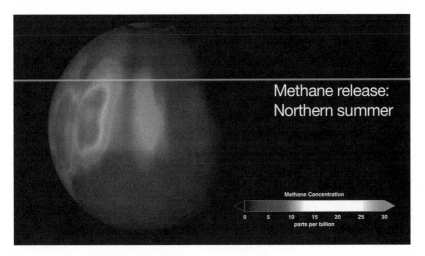

Figure 14.1. Map of multiple plumes of methane, seen in Martian summer in the northern hemisphere in 2003, as reported by Mumma et al. (*Science*, 2009). The methane appears to be present globally, but is strongest in limited locations. See Plate 6. Image courtesy of NASA.

(about 14 parts per billion) and south (less than about 6 parts per billion) of 40° latitude and rose as high as about 24 parts per billion at equatorial latitudes. From the error bars that are included visually in a graph, we find that it is a rare measurement that by itself exceeds three times the noise level, which is the widely accepted threshold for believability for detections. For example, the methane measurements at –60° (near the South pole of Mars) appear to be 4 ± 3.5 parts per billion (for the "R0" line) and 8.5 ± 3.5 parts per billion (for the "R1" line). The first of these measurements is effectively nothing but noise; the second is a possible, but marginal, detection. Neither of those results is, statistically, a definitive detection. Both are consistent with a measured value of zero (or more precisely, an upper limit (99.7 percent confidence level) of 10.5 parts per billion in both cases). Thirty degrees north of the equator, the two measurements are 16 ± 5 parts per billion (for the "R0" line) and 27 ± 7 parts per billion (for the "R1" line). These two results are both greater than three times the noise and so they might be significant, but they are nevertheless marginal, not overwhelming, stop-the-presses detections. To Mumma's credit, though, he reported detections in two rather than just one line, which adds a level of credibility to his results that most others, with detections in but a single line, lack.

By 2009, the Mumma team's earlier claim, made in 2006, for a signal as strong as 250 parts per billion at equatorial latitudes, the result that "cannot be waved away as a measurement error," had evaporated. The earlier claim for 60 parts per billion? That also was gone. The strongest signal reported in the 2009 paper was 29 ± 9 parts per billion in a single line and the strongest "mean" value (the average of two different methane lines at a single position on Mars) was 24 parts per billion. Mumma argued, similarly to the argument put forward by Krasnopolsky in 2004, that the data were consistent with "two local source regions," each one releasing massive plumes of methane for a few months. These plumes rise into the atmosphere and then rapidly disperse, such that the methane becomes undetectable. At a press conference staged by NASA in 2009 at which Mumma announced these results, others mentioned the possibility that the source of the methane might be Martian bacteria. Mumma, himself, did not make that claim; however, unlike in 2004, he made no attempt to dissuade listeners of that possibility.

Mumma did tell a reporter for the *New York Times*, "This is the first definitive detection of methane on Mars."[10] Keep in mind that he is reporting his results in 2009 and both Krasnopolsky and the Mars Express team claimed similar detections of methane on Mars in 2004 and published their results in 2004. If any of the three detections are correct, then all three are probably correct. If any of them are wrong, they all are probably wrong. Whether Mumma's work is right or wrong scientifically, his claim of priority for making the first definitive detection of methane on Mars was historically incorrect. At best, that scientific first belongs to or is shared with others, but the stakes are very high.

Remarkably, by January 2006 the methane apparently detected in 2003 and early 2004 by three different observing teams was effectively gone. Mumma had observed Mars using the detector called CSHELL on the IRTF, on January 26, 2006, and he reported these results, along with his 2003 measurements, in his 2009 paper. Near the Mars equator, Mumma reported a measurement of about 4 parts per billion of methane (the error on these measurements appears to

be about 2 parts per billion) in 2006. At more northerly and southerly latitudes, the 2006 methane level he presented is at about 1–2
parts per billion.[11] Rather than identify these measurements as upper
limits (plausibly, less than 6 parts per billion) and nondetections, he
chose to refer to them as revealing a "low mean abundance" of 3
parts per billion of methane on Mars. Whether the methane levels in
2006 were effectively zero or 3 parts per billion, they were down pre-
cipitously from the levels of 20 and 250 and 14 and 24 and 29 parts
per billion he had previously reported for the 2003 measurements.
Either the methane had mostly disappeared, had completely disappeared, or had never existed in the first place.

The existence of apparently detectable levels of methane in 2003
and 2004 combined with the absence of detectable methane by early
2006 implies a destruction lifetime of less than three Earth years for
methane in the atmosphere of Mars. Three years is about one hundred times faster than the 350-year lifetime predicted by atmospheric
models for the destruction of methane in the Martian atmosphere.
As if the entire Martian methane saga were not already odd enough,
the 2006 results rewrote the script entirely. Not only do we need one
or more sources for the methane that are confined to very small, localized regions of the Martian surface and that can belch out significant amounts of methane in very short amounts of time, we now
also need a process that can remove atmospheric methane one hundred times more efficiently than can sunlight and photochemical
reactions.

Interestingly, a few years later the Mars Express team would offer
support for Mumma's idea that methane in the atmosphere of Mars
may have a very short lifetime; however, their conclusions were also
drawn from a set of measurements of Martian methane that claim
high levels of methane at the north pole and low levels of methane
at the equator, which is the reverse of what Mumma found. In 2011,
Geminale, Formisano, and their colleague G. Sindoni revisited the
issue of the lifetime of atmospheric Martian methane, as discovered
in the Mars Express data.[12] As they noted, with a lifetime of 300–600
years, the methane should be geographically well mixed in the atmosphere, and Mars Express should not have been able to detect spatial
and seasonal changes in the level of methane in the atmosphere of

Mars. Yet it did detect such changes. Of greatest interest, "the geo-graphic map in northern summer shows a methane abundance peak [of 45 parts per billion] over the north pole when the polar ice is sub-limating," which implicates the north polar ice cap as an important source of Martian methane. In addition, they concluded that "one or more strong destruction mechanisms and/or [a] strong sources are contributing to the not uniform distribution of methane in the atmosphere." According to their calculations, given the amount of methane (averaging about 8,700 tons) released annually from the north polar region, the lifetime of methane in the atmosphere is only about 12 years. Taking into account the seasonal variation of two to three in the amount of atmospheric methane seen from northern spring to northern winter, they suggest the atmospheric lifetime of Martian methane might be as short as 4 to 6 years. They suggest that near the surface, oxygen-rich materials could react with and destroy the methane; in addition, they hypothesize that high-energy elec-trons that are generated during seasonal Martian dust storms may be efficient at destroying methane molecules.

Meanwhile, while Mumma chased Krasnopolsky, Krasnopolsky continued to chase Mumma. Krasnopolsky made his next round of observations on February 10, 2006, using the same CSHELL detector on the same telescope, the IRTF, as Mumma had used just two weeks earlier, on January 26. This time, Krasnopolsky did not detect any methane. He reported an upper limit of 14 parts per billion.[13] This nondetection, Krasnopolsky points out, "does not rule out the value of 10 ppb" observed by his team from the Canada-France-Hawai'i Tele-scope in 1999 or by the Mars Express team in 2004. It should, how-ever, make impartial adjudicators once again question the accuracy of the 70 ± 50 parts per billion measurement made in 1988, just in case anyone still believed that result. This nondetection by Krasnopolsky in 2006 is also consistent with the results reported by Mumma (in 2009) for January 2006 as a "low mean abundance," rather than as a nondetection—using the same detector on the same telescope on the same mountain. Thus, on one point all observers appeared to agree: in early 2006, the amount of methane in the Martian atmosphere was either nonexistent or so low as to be at or below the threshold of de-tectability. Even that single point of agreement would not last.

In 2012, Krasnopolsky published a reanalysis of his 2006 CHSELL/ IRTF data, combined with additional observations he made with the same detector and telescope in December 2009. He applied a large number of new corrections to the data. One of these corrections was for the amount of scattered light that can contaminate a spectrum. When observing Mars from a telescope on Earth, scattered light from external sources like the Moon, the lights of the city of Hilo at the base of Mauna Kea, headlights of cars driving on the roads on the big island of Hawai'i, small LEDs on computer screens in the telescope dome, as well as scattered light internal to the detector system itself can contaminate the data. The scattered light has to be identified and removed from the data, as best as is possible. Other corrections have to be made for the effects of water vapor, ozone, and methane in Earth's atmosphere. These molecules, which are in the air above the telescope and which are very sensitive to the pressure and tempera- ture in Earth's atmosphere at the times of the measurements, will all impact the very fine details of the spectrum made from the light from Mars that had to pass through Earth's atmosphere in order to reach the telescope on the summit of Mauna Kea.

In 2012, Krasnopolsky claimed to have improved his ability to make all the needed corrections in his processing of the 2006 ob- servations. As a result, he was able to substantially lower the level of noise he inferred to exist in those data. With his improved ability to model and remove the noise in his data, he revised his analysis of his 2006 CHSELL observations. His new conclusion: during what Krasnopolsky now claimed were perfect observing conditions,[14] the data show that methane was detected at stunningly low levels in three distinct regions on Mars. In the south, from 80° south to as far northward as 45° south, the data show an average of 1.7 ± 1.2 parts per billion of methane. Krasnopolsky claimed this result as a detec- tion, though others might see these results as only noise, reporting instead a measured value of less than 3.6 parts per billion. Closer to the equator, from 45° south to 7° north, the methane levels averaged 10 ± 2 parts per billion, while from 7° north to 55° north, the meth- ane levels averaged 4.4 ± 1.2 parts per billion. If all the steps in Kras- nopolsky's analysis of the 2006 data set are correct, then at least the 10 ± 2 result near the Martian equator is a solid detection of methane

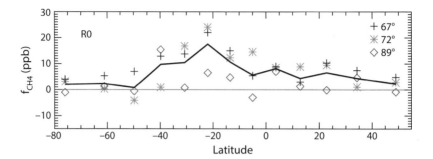

Figure 14.2. February 2006 measurements, showing the abundance of methane in the atmosphere (in parts per billion) as a function of Martian latitude, from –80 degrees (south) to +60 degrees (north). Measurements were made at three different longitudes on Mars (67W, 72W, 89W), as indicated by the three different symbols, and were made at the particular methane absorption line known as the R0 line (at 3028.752 inverse centimeters). The data appear to show a detectable amount of methane at a level of 10–20 ppb) at mid-southern latitudes and lower amounts of methane, or no methane, farther north and south. Image based on Krasnopolsky et al., *Icarus*, 2012.

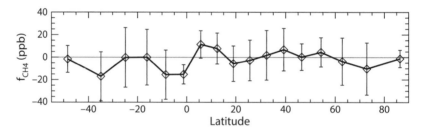

Figure 14.3. December 2009 measurements, showing the abundance of methane in the atmosphere (in parts per billion) as a function of Martian latitude, from –60 degrees (south) to +90 degrees (north). No methane is seen (solid horizontal line indicates mean value of zero; solid vertical lines indicate standard deviations on the measured values). Image based on Krasnopolsky et al., *Icarus*, 2012.

on Mars. The greatest methane abundance, he reported, overlaps the deep canyons of the great Martian canyon system, Valles Marineris. Krasnopolsky found reassurance in the fact that these results are, in his view, in agreement with the February 26, 2006, measurements reported by Mumma's team (in 2009) of only about 3 parts per billion. Others might conclude that Krasnopolsky's new results contradict Mumma's results; perhaps Krasnopolsky over-modeled his data and pulled signal out of the noise where there is, in fact, no signal.

Just three years after his 2006 IRTF observations, in a data set obtained in 2009, again using the IRTF and, when Mars was less favorably positioned than in February 2006, Krasnopolsky did not detect methane in the Martian atmosphere. The upper limits in the 2009 data, based on his analysis, were 15 parts per billion for one line and 8 parts per billion for another.

Can we draw any conclusions from Krasnopolsky's work? Yes. Mars either has very little methane or no methane at all. Mars showed no detectable levels of methane in 2009. But with the noise level in his data in 2009, he would have been unable to detect methane at a level of 1.7, 4.4, or even 10 parts per billion, so his nondetections in 2009 are consistent with his claimed detections from 2006.

Krasnopolsky and Mumma independently measured the methane abundance on Mars in January and February 2006 with the same detector on the same telescope on the same mountain and initially reported the same answer: Mars showed no, or almost no, detectable methane. That is, they agreed on the answer until Krasnopolsky revisited his data and got a different answer. Given the very low "signal" values in his revised results, the more likely answer is that Mars showed no detectable levels of methane in 2006, though perhaps the amount of methane in both years was real, but extraordinarily small.

We are left with several methane mysteries. Why did three different groups detect very low levels of methane on Mars in 2003 and 2004? How did that methane become so depleted, or even disappear, in just three years? Or was all of it noise? Perhaps methane was never present on Mars in the first place.

mars global surveyor

While the ground-based interpretations of Martian methane spectra were descending into controversy, new and independent confirmation of the presence of methane on Mars appeared to come from measurements made from Martian orbit by the Thermal Emission Spectrometer (TES) on the Mars Global Surveyor (MGS) spacecraft, as reported by S. Fonti and G. A. Marzo in 2010.[15] (Recall that ESA's Mars Express had—perhaps—already detected methane from

Martian orbit in 2003.) MGS was launched in 1996 and became the first orbiting mission around Mars since the Viking missions in the 1970s. Once MGS settled into orbit, it began studying Mars and continues to do so. The TES was not designed to be sensitive enough to detect the narrow lines of methane that might be present in spectra of the Martian atmosphere; however, Fonti and Marzo used a combination of computational and experimental techniques to show, they claimed, that they could, in fact, detect methane with TES. In addition, TES was sensitive to the methane absorption band at a wavelength of 7.7 microns rather than at 3.3 microns. This band is the second strongest band for methane, after the 3.3-micron band, but no ground-based telescope can see through Earth's atmosphere at this wavelength. Only a Mars-orbiting spacecraft with a telescope looking down at Mars could study the Martian atmosphere and search for signs of methane absorption at 7.7 microns. TES therefore had the potential to make a major contribution to our understanding of whether methane exists in the Martian atmosphere.

The technique they employed, called "cluster analysis," had been used previously on other TES data. Cluster analysis, they claimed, ultimately allowed them to carefully select only the data deemed statistically reliable for methane; that is, of the nearly three million spectra obtained by TES that cover the methane band, fully 59–86 percent (depending on the year of the observations) were rejected. Any technique for selecting or rejecting data is fraught with risk. Those who use the cluster analysis technique think that it is very robust at rejecting data negatively impacted by systematic errors without also rejecting data because of random noise. More skeptical readers might suggest that this technique is a good way to reject only the data you don't want in order to produce a result that you do want.

According to Fonti and Marzo, once the Martian spectral data from TES were carefully culled, with the supposedly bad data removed, TES was able to detect methane, including spatial and seasonal variations in the methane abundance. We can draw one of two conclusions from these results: (a) if you use an algorithm to carefully select only the data that give the results you want, you get the results you desire; or (b) if you judiciously and objectively find a way to throw out bad data, the good data that remain are impressively good.

Figure 14.4. Mars Global Surveyor spacecraft, as imagined in orbit above Mars. See Plate 7. Image courtesy of NASA/JPL-Caltech.

In their measurements, which they published in 2010, the global methane abundance in the northern hemisphere was high at the time of every Martian autumnal equinox (33 ± 9 parts per billion in 1998–2000; 18 ± 7 parts per billion in 2000–2002; 30 ± 8 parts per billion in 2002–2004) and low at the times of the winter solstice (6 ± 2, 5 ± 2, and 5 ± 1 parts per billion in these same three Martian epochs,

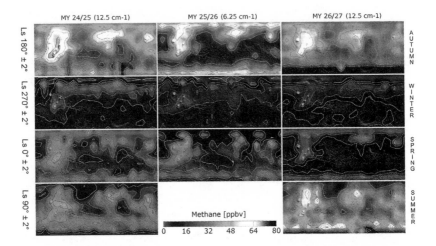

Figure 14.5. Maps of the spatial distribution of the apparent abundance of methane on Mars, showing differences in methane abundance with longitude and season, as calculated from Mars Global Surveyor data. Each map extends in latitude from –60 degrees (south) to +60 degrees (north). In each Martian year (vertical columns for MY 24/25, MY 25/26 and MY 26/27) maps are shown for each Martian season. For each Martian season, a longitudinal slice is presented for each of the three years. For Martian autumn, the longitude near 180 degrees is shown; for Martian winter, the longitude centered on 270 degrees is shown; for Martian spring, 0 degrees, for Martian summer, 90 degrees. Grayscale indicates the abundance of methane, from a high of about 80 parts per billion (white) to a low of a few parts per billion (dark). At all seasons, more methane is seen at latitudes of 90 and 180 degrees than at 0 and 270 degrees. And more methane is seen in summer and autumn than in winter and spring in all three Martian years. See Plate 8. Image from Fonti and Marzo, *Astronomy & Astrophysics* 2010; reproduced with permission from *Astronomy & Astrophysics*, © ESO.

respectively). Their maps identified "three broad regions where the methane amount is systematically higher (Tharsis, Arabia Terrae, and Elysium)." Tharsis and Elysium are volcanic regions, while Arabia Terrae is a very high-elevation region that is densely cratered. All in all, their results appeared consistent with those reported by some other groups; in particular, the phenomenon of spatial variations in the methane abundance across the Martian planet as well as the short temporal scale for the lifetime of methane in the Martian atmosphere (about 0.6 years, in their estimation) seemed to replicate the big-picture aspects of the measurements by the Krasnopolsky-led, Mars Express, and Mumma-led teams.

The MGS/TES story of methane, however, did not end with this initial analysis. In 2015, Fonti and a team of six collaborators (not

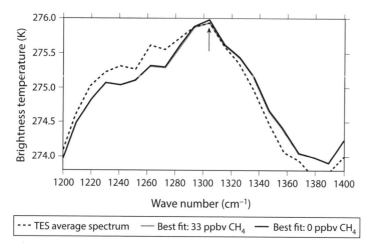

Figure 14.6. Plot of amount of reflected light from Mars (brightness temperature) versus wavelength (wavenumber) across the narrow region at which methane should impact the spectrum (at 1304 inverse centimeters). The dashed line represents the data. Two model fits to the data are also plotted, a black line (for a Martian atmosphere that contains no methane) and a gray line (for a Martian atmosphere that contains 33 parts per billion of methane). Except for a single data point at the position 1304, the black and gray lines are perfectly superposed. While the gray (with methane) line is ever so slightly a better fit to the data at 1304 inverse centimeters, the discrepancy between the data and the models across most of the spectrum is much larger than the difference between the two models, which suggests that the precision needed to definitely identify the presence of methane in the Martian atmosphere, is, in the words of the authors of this research, "extremely hard to reach, if even impossible, at present." Therefore, the conclusions drawn by Fonti (2010), as shown in figure 14.5, that purportedly reveal methane in the Martian atmosphere, are withdrawn, as those results cannot be confirmed or refuted. Image from Fonti et al., *Astronomy & Astrophysics* 2015. Reproduced with permission from *Astronomy & Astrophysics*, © ESO.

including Marzo) revisited the earlier cluster analysis of the Mars Global Surveyor TES data.[16] After expending a tremendous effort at reproducing and improving on the original analysis of Fonti and Marzo, Fonti and his team came to a profound realization: "Unfortunately, the conclusion of this effort is that, in spite of all attempts, we have been unable to produce distinct methane and methane-free clusters and, as a result, we cannot use the same method as Fonti and Marzo (2010) to estimate the methane abundance." That is, the Fonti and Marzo method and the numbers derived therefrom are inherently and fundamentally untrustworthy. Fonti's 2015 team of collaborators generated synthetic spectra to compare to their observed spectra and found that they could not distinguish between models that included 33 parts per billion of methane and models that had absolutely no methane. The conclusion of their work is that they could not unambiguously identify methane in the TES data. The original design for Mars Global Surveyor, in which it did not have the ability to detect methane on Mars, was correct. TES did not and cannot provide independent support for or against the claims made by Krasnopolsky, Mars Express, and Mumma for detections of low levels of methane in the Martian atmosphere in the first few years of the twenty-first century.

methane escapes detection in multi-telescope study

Geronimo Villanueva, of NASA's Goddard Space Flight Center, teamed up with Mike Mumma and an international team of eight other astronomers to carry out a multiple-year study of Mars using several Earth-based telescopes. Villanueva is a highly respected young planetary scientist, who was awarded the Urey Prize (as the young planetary scientist of the year) by the American Astronomical Society in 2015. He was also a member of Mike Mumma's team that reported (in 2009) detecting plumes of methane in the Martian atmosphere in 2003. Villanueva's team was, at least in part, motivated by "the recent observations of methane by four groups [that] indicate regions of localized release, and high temporal variability. The complexity of the observations have prompted questions of the reliability of the detections."[17] One of these four groups, of course, the Mars Global Surveyor team, would later retract their claim of having detected methane.

Villanueva put together a team to use several telescopes on Earth to search for minor atmospheric constituents in the Martian atmosphere. His team cast a wide net in their study of Mars. In addition to methane, they were attempting to identify a whole host of carbon-containing gases (e.g., C_2H_2 [acetylene], C_2H_4 [ethylene], C_2H_6 [ethane], H_2CO [formaldehyde], CH_3OH [methanol]), nitrogen-containing compounds (N_2O [nitrous oxide], NH_3 [ammonia], HCN [hydrogen cyanide],), and chlorine-containing species (HCl [hydrogen chloride], CH_3Cl [methyl chloride]). They invested four years of effort, from 2006 through 2010, using three powerful telescopes: the Very Large Telescope (VLT) on Cerro Paranal in northern Chile, which is a set of four 8-meter (26-foot) telescopes designed to work together; the 10-meter Keck-2 telescope on Mauna Kea; and the IRTF telescope, also on Mauna Kea, which is "only" a 3-meter telescope.

It so happened that Villanueva observed Mars with the same telescope (the IRTF), with the same detector system (CSHELL), "in the same season" and "over the same region" of Mars as did Krasnopolsky. In fact, they did the same experiment about one month apart: Villanueva studied Mars on January 6, 2006 (Mumma first reported previously on these same observations in his 2009 paper), while Krasnopolsky studied Mars four weeks later, on February 2. The surprise and puzzle of these twin observing projects is that Villanueva and Krasnopolsky obtained different answers.

Villanueva published his team's research results in the journal *Icarus*, in a paper that is now one of the most widely read and cited articles in all of planetary sciences for the year 2013.[18] He did not find any methane, with "three-sigma" (99.7 percent confidence level) detection thresholds of 7.8 parts per billion on January 6, 2006; 6.6 parts per billion on November 20, 2009; and 7.2 parts per billion on April 28, 2010.

The Mumma-led team (including Villanueva) had previously reported results for the January 2006 observing run of between 1 and 6 parts per billion for three different latitude regions on Mars. The Villanueva-led (including Mumma) team's answer in 2013 and the Mumma-led (including Villanueva) team's answer in 2009, using the same data, can therefore be understood as consistent, though the Mumma-led team in 2009 clearly believed that they had detected methane, not simply measured an upper limit for the possible amount of methane present.

Krasnopolsky first reported an upper limit for his February 2006 measurements of less than 14 parts per billion, but later corrected his result and reported firm detections ranging from 1.7 to 10 parts per billion, depending on the location on Mars. If Villanueva's reanalysis of the January 2006 data is correct and if Krasnopolsky's reanalysis

of the February 2006 measurements is correct, Mars must have released a tremendous amount of methane over the last three weeks
of January, so that the methane concentration in the atmosphere of
Mars had increased from an undetectable level in early January to
detectable levels in early February, when Krasnopolsky showed up at
the telescope. Villanueva calculated that the amount of methane that
would have been pumped into the Martian atmosphere over those
three weeks amounts to 4,500 metric tons in 27 days or less; this
much methane implies a mean rate of 2 kilograms (4 pounds) per
second of methane expelled continuously into the atmosphere for
27 days from a subsurface reservoir. This rate would be, by any measure, "quite extraordinary." If Krasnopolsky's measurements were not
already suspect, Villanueva's results cast into very strong doubt the
reliability of Krasnopolsky's revised "detections" reported in 2012.

Villanueva, Mumma, and their colleagues also calculated that the
enormous "principal plume" of methane that must have released the
methane purportedly observed by several groups in 2003 (and perhaps 2004) "contained ~19,000 metric tons of methane." They went
on to say that this volume of methane is "comparable to that of the
massive hydrocarbon seep at Coal Oil Point (Santa Barbara, CA)."
They then pointed out that this Martian methane must have disappeared over three years' time: "By January 2006, the global abundance of methane represented only about 50 percent of the quantity
released in March 2003, suggesting rapid destruction." If the Villanueva measurement for the methane abundance in January 2006
(effectively zero methane in the atmosphere) is correct, and if rapid
sinks for methane do not exist, these results also would seem to call
into question the accuracy of the "detections" reported by multiple
groups for the year 2003, although Villanueva, Mumma, and their
co-authors of the 2013 paper don't actually do that.

The rapid destruction of methane is problematic. Quite a few atmospheric scientists question whether destructive mechanisms for
that much methane can act that quickly; however, a short lifetime for
methane in the atmosphere of Mars, such as the process suggested
by the Mars Express team in 2011, appears to be the only plausible
explanation that might allow most or all of the reported methane
observations to be correct.

How does the Villanueva team make heads or tails out of these conflicting results? "Considering the complexities associated with measuring methane at the low spectral resolutions featured by MGS/TES and PFS/MEX," they wrote, "those measurements may be affected by instrumental effects (e.g., micro-vibrations) and unrecognized solar/water signatures that dominate the spectral regions used by these teams to search for methane." In other words, the "detections" claimed by these orbiting spacecraft are likely wrong. As we know now, Villanueva was correct that the detection of methane reported by MGS/TES was wrong. He might also be right about the PFS/MES results.

The Villanueva team concluded, "If methane is being released into the atmosphere, this process is probably sporadic and not continuous." That "if" is a very big "if."

15

—

sniffing martian air
with curiosity

NASA entered the Mars rover business with the launch of the Pathfinder mission in 1997. Pathfinder landed on Mars and rolled out Sojourner, a 23-pound (10.6 kilogram) device about 2 feet long, 1.5 feet wide, and 1 foot tall. The size of a microwave oven, Sojourner puttered around the surface of Mars at a speed of about 1.5 feet per minute (or about 1 mile in 59 hours).[1] In comparison, box turtles can move at top speeds of about one-quarter mile per hour (but they don't maintain that speed for long). Sojourner survived for 83 days (and only traveled a total distance of about 330 feet during its active lifetime) before running out of power. As an engineering experiment, Sojourner was a great success. NASA proved that it had the wherewithal to land robotic rovers on Mars. As a science mission, Sojourner had little impact, but engineering, not science, was the goal for this rover. The next step for NASA was to make bigger, better rovers capable of doing significant science.

Sojourner's successors were the Spirit and Opportunity rovers, which were launched in June and July 2003. Spirit and Opportunity each weigh 384 pounds (174 kilograms) and are 5.2 feet long and 4.9 feet tall. Spirit covered almost 5 miles before sending its last communication to Earth in March 2010. Opportunity has traveled more than 26 miles across the Martian surface and, after 15 years, remains active.

As engineering experiments, Spirit and Opportunity were tremendously successful; as geology experiments, Spirit and Opportunity have enabled scientific accomplishments far beyond what mission scientists hoped to have achieved. Neither Spirit nor Opportunity had instruments that could look for methane in the Martian atmosphere; however, the scientific and engineering knowledge gained from the work of these rovers paved the way for NASA's next rover mission, which was designed to search for Martian methane.

NASA launched the Mars Science Laboratory Curiosity rover on its journey to Mars in November 2011. After arriving at Mars, the spacecraft descended on a parachute toward the Martian surface. On August 5, 2012, rockets fired seconds before landing, allowing the spacecraft to hover above Gale Crater and use a tether to lower Curiosity onto the surface of Mars inside the crater. Once the landing system had safely deposited the rover on its wheels, NASA cut the tether and sent the landing system off to crash into the Martian surface.

The six-wheeled, 1,982-pound Curiosity is 10 feet in length, 9 feet in width, 7 feet tall, and has a top speed of 90 inches (7.5 feet) per minute. Curiosity is a smart, scientifically sensitive, extremely well equipped, and robotically capable scientific laboratory slowly wandering around the surface of Mars. The Tunable Laser Spectrometer (TLS) is part of the Sample Analysis at Mars (SAM) instrument suite included as part of Curiosity's package of scientific devices. Using TLS, SAM was designed to allow Curiosity to detect methane and end five decades of controversy regarding methane on Mars.

Being on the surface of Mars, SAM has enormous advantages over telescopes on Earth or even telescopes in orbit around Mars. The intake nozzle for SAM is located only about 3 feet above the Martian surface, so it is sampling gases that are present in the lowest part of the Martian atmosphere. Also, the spectral resolution of SAM, that is, its ability to see individual lines of a gas like methane, is "far superior to those of the ground-based telescopic and orbiting spectrometers" and offers "unambiguous identification of methane in a distinct fingerprint spectral pattern of three" (rather than just one) different lines in the 3.3-micron region of the near-infrared spectrum of methane.[2]

SAM sucks Martian air into an 8-inch-long (21-centimeters-long) sealed chamber, called a Herriott cell. Then, SAM bounces a laser

Figure 15.1. Mars Curiosity rover self-portrait, obtained on May 11, 2016, at a drilled sample site called "Okoruso," on the "Naukluft Plateau" of lower Mount Sharp. An upper portion of Mount Sharp is prominent on the horizon. For scale, the rover's wheels are 20 inches (50 centimeters) in diameter and about 16 inches (40 centimeters) wide. See Plate 9. Image courtesy of NASA/JPL.

pulse back and forth—through a total path length of 55.1 feet (16.8 meters)—through the captured Martian gas. The laser source, along with a handful of specially designed mirrors that the scientists use to steer the laser pulse into the Herriott cell, is in a device called the foreoptics chamber that is attached to the Herriott cell, so the laser light passes first, just one time, through the foreoptics chamber and then reflects eighty times within the Herriott cell before the detector measures the light.

This mini-experiment is repeated, over and over again in 2-minute intervals, for one hour to complete one set of measurements. By comparing the average laser signal through a chamber full of Martian air (a "full cell") with the average laser signal when passed through the chamber that has been evacuated (the "empty cell"), SAM scientists can measure the properties of the light signal and thereby determine the contents of the gas, because the gas molecules leave their imprint on the light spectrum. If the Martian atmosphere contains methane, SAM should be able to detect that methane down to a phenomenally low level of a few hundred parts per trillion (or a few tenths of a part per billion), assuming no contamination occurs and no systematic problems develop with the apparatus. If methane exists in the Martian atmosphere at the level of even a few parts per billion, SAM and the TLS should detect it with ease. That, at least, is how the experiment was designed to work.

A SAM experiment was run in 2013 to measure the methane abundance on six different nonconsecutive Martian sols (one sol is one day on Mars, equal to a period of 24 hours, 39 minutes, 35.244 seconds). Those six sols were spaced across a span of 234 Martian sols, beginning in October 2012 (sol 79) and continuing through June 2013 (sol 313). Three of these days fell in Martian spring in the southern hemisphere of Mars; three of these days fell in mid-late southern Martian summer.

In October 2013, Christopher Webster of the Jet Propulsion Lab, Paul Mahaffy of NASA's Goddard Space Flight Center and the Principal Investigator of SAM, and the entire Mars Space Laboratory Science Team reported on these six sols of measurements in the journal *Science*, "To date, we have no detection of methane."[3] No equivocation, no uncertainty, and no methane.

No methane was detected during any of the six individual Martian sols during the study. They reported that the average signal was an incredibly low 0.18 ± 0.67 parts per billion. At a 95 percent confidence level (twice the error level), the methane abundance was below 1.3 parts per billion; at a 99.7 percent confidence level, the methane abundance was below 2 parts per billion, "which reduces the probability of current methanogenic microbial activity on Mars and limits the recent contribution from extraplanetary and geologic sources." This experiment, having been done on the surface of Mars, required no assumption-driven computer modeling about how much methane is in the atmosphere of Earth and no other assumptions about other sources of noise and interference generated by sources of emission and absorption in Earth's atmosphere.

Were the methane abundance in the Martian atmosphere 250 or 70 or 30 or even 8 parts per billion, detecting Martian methane would have been child's play for SAM. Could SAM have failed to detect the methane, were the methane at any of those levels? No. Checkmate. Game over. No methane on Mars. Except, as twentieth-century philosopher and New York Yankee Hall of Fame catcher and manager Yogi Berra used to say, "It ain't over 'til it's over." And in this saga of Martian methane, Yogi Berra would be right.

an extraordinary claim

In December 2015, with more data in hand, Webster, Mahaffy, and the SAM science team reported a different answer. This time, they announced that they had "observed elevated levels of methane" over a two-month interval from November 29, 2013 (sol 466) through January 28, 2014 (sol 526). They still reported undetectable levels of methane during the six sols of observations in 2012 and 2013 on which they originally reported in their first paper on Curiosity results for Martian methane; they also reported undetectable levels of methane for a later period, from March 17, 2014 (sol 573) through July 9, 2014 (sol 684). In that sense, they did not change their interpretation of their original data set, although they did improve their analysis and revise their numerical results. Their measurements from late 2013 and early 2014, however, appear to tell a dramatic story of the brief appearance of methane, a time period when a Martian hiccup may have expelled some methane into the atmosphere of Mars.[4]

In order to understand the SAM measurements and decide whether the extraordinary claim by the SAM team for detecting Martian methane is correct, we need to know that before the launch of Curiosity from Florida, the experimenters injected air with a known abundance of methane, at a level of 88 ± 0.5 parts per billion, into the Herriott cell. This gas was used to calibrate the future TLS measurements that would be made on Mars; presumably, all of this air was pumped back out of the Herriott cell before launch. Unfortunately, ambient Florida air apparently leaked into the foreoptics chamber, so that both "empty-cell" and "full-cell" measurements detected 10 parts per million (a mind-blowing 10,000 parts per billion) of Florida methane upon arrival on Mars. Either some of the methane from the "known abundance" test remained in the Herriott cell, or some Florida air leaked from the foreoptics chamber into the Herriott cell, or both. All measurements through the first 78 Martian sols were discarded due to presumed contamination by terrestrial air. After using a vacuum pump to attempt to remove the terrestrial air from the foreoptics chamber, all measurements from sol 79 through sol 292 showed about 90 parts per billion methane both when the Herriott cell was "empty" and when it was "full," which means the foreoptics chamber still contained a large amount of Florida methane, even after nearly a year on the surface of Mars. Apparently, despite the best efforts of the SAM team, they were unable to successfully pump all the methane out of their experimental device.

Things did get better, however. In experiments run on sols 306 and 313, they apparently had reduced the methane contents to less than 20 parts per billion in both empty-cell and full-cell measurements. Though this is, indeed, an impressive improvement, this means the amount of Florida methane remaining in the foreoptics chamber on both of these sols was still enormous in comparison to the expected methane abundance in Martian air. The bottom line is that, in all of the SAM measurements of methane on Mars, all of their spectral measurements, whether empty cell or full cell, are dominated by the contribution of the substantial amount of terrestrial methane in the foreoptics chamber. Ultimately, despite having their detector system on the surface of Mars, the Curiosity team still had the same problem that observers do when looking through Earth's atmosphere to study Mars.

In principle, if the experiment works perfectly, this very large methane signal is subtracted completely from the signal of Martian methane. We also need to remember, however, that all claims of measuring the presence or absence of Martian methane from Curiosity involve subtracting one very big number (the amount of methane detected in the empty-cell measurements) from another very big number (the amount of methane detected in the full-cell measurements) in order to accurately measure a very small number. Can the SAM team do this successfully at the level of a fraction of a part per billion?*

In the January 2015 report, Webster and his team had data from 13 rather than just 6 Martian sols. With more data in hand, they reanalyzed their entire data set. Six of the 13 sols were considered "low-methane" runs; four were identified as "high-methane" runs; two were test runs. One run had a very high background and so was not included in the final analysis.

Again, they reported that the background level of methane in the Mars atmosphere was quite low. The mean value was 0.69 ± 0.25 parts per billion. This average value reported in 2015 was a bit higher (0.69 versus 0.18) and the level of the noise a bit lower (0.25 versus 0.67) than the values reported in 2013, but the mean value was still consistent with a null result or a phenomenally low level of methane. Put simply, from sol 79 through sol 313 and again from sol 573 through 684, Mars had much less than 1 part per billion of methane in its atmosphere. This is as close to zero methane as TLS can measure.

Something mysterious happened, however, during a 60 Martian-sol interval, extending from November 29, 2013 (sol 466) until

---

*The mathematical analogy to this measurement is that the experimenters are making two measurements, one of X+Y and the other of X. They then subtract X from X+Y, i.e., $(X+Y) - X = Y$. If the value of X is measured to be identical in both measurements, the accuracy with which the value of Y is known is very high; however, if the size of the errors on the measurements of $X + Y$ and of X are comparable to the magnitude of Y, which is likely if the value of X is thousands or tens of thousands of times greater than Y, you would be justified in having limited confidence in your knowledge of the value of Y.

January 28, 2014 (sol 526). According to the SAM measurements, on sol 466 the level of methane had risen to a level ten times above the background level of 0.69 parts per billion, averaging about 7.2 ± 2.1 parts per billion in measurements made on sols 466 (5.48 ± 2.19 ppb), 474 (6.88 ± 2.11 ppb), 504 (6.91 ± 1.84 ppb), and 526 (9.34 ± 2.16 ppb). The experimenters could not determine whether the methane level rose slowly over a period of months or quickly over a period of weeks, days, or hours. Then, in less than 47 sols (by the next measurement day on sol 573), the methane level had dropped suddenly and dramatically to well below 1 part per billion (0.47 ± 0.11 ppb).

No one questions the ability of SAM to detect methane if it is present. If SAM detected methane inside the Herriott cell, then methane was present inside the Herriott cell. Somehow, during the 153-sol period from sol 313 to sol 466, the level of methane in the environment of the SAM detector increased dramatically. Then, over a similar period of time, from sol 526 to sol 573, some process removed the methane. The big question is: What happened?

Could the methane detected by SAM be due to terrestrial contamination? Could the methane be due to some other problem in the analysis performed by the Curiosity Science Team? The Science Team acknowledged that "an undetected analytical problem cannot be ruled out," but they also "conclude that the possibility of terrestrial contamination producing the 'high' methane signals is very unlikely."

If real, how could the methane have disappeared so quickly? Could Mars produce and then remove that much methane in just a few weeks? If so, Mars must have a sink, a black hole, a vacuum, a destructive process of unprecedented power that destroys the gas as quickly as the planet produces it. The sink must operate on a time frame of days to weeks, not 3–6 years or even 300 years. Such a rapid process for removing methane from the atmosphere is unlikely, and if the methane is not destroyed overnight, another explanation is needed.

According to Webster, writing for the SAM Science Team, these results imply "that Mars is episodically producing methane from an additional unknown source." In addition, the source of methane almost certainly must be local, not widespread, across Mars, and

Curiosity was fortunate to be in the location where Mars burped out methane. Furthermore, as soon as Mars belched out methane in the vicinity of Curiosity, Martian winds quickly dispersed this methane plume far and wide such that the methane became undetectable by Curiosity a few weeks later. Last, the methane source itself must have been short-lived, as after dispersal of the original methane plume the methane source did not replenish the local atmospheric supply to detectable levels.

These dramatic results were reported in *Science* by a large, international team of extremely highly respected scientists participating in a billion-dollar NASA mission. Their work was vetted by other scientist-referees and puzzled over by skeptical magazine editors before publication. These results are extraordinary, and extraordinary results demand a high level of proof. Can such an extraordinary claim be right?

Even the editors of *Science* seemed to want to offer their readers a dose of skepticism. They commissioned a "perspectives" editorial that they published in the same January 23, 2015, issue in which they published the Webster team claim that the Curiosity team had detected Martian methane. The author of the perspectives piece, Kevin Zahnle, questioned whether the Mars Curiosity TLS team had detected *Martian* methane. Perhaps, he suggested, Webster and Mahaffy had detected methane gas from Earth that they had brought with them from Earth.

Zahnle, a planetary scientist at NASA's Ames Research Center and a fellow of the American Geophysical Union, has regularly challenged the claims that detections of Martian methane are correct. He fondly quotes Carl Sagan, one of the great supporters of the search for life on Mars, who said, "extraordinary claims require extraordinary evidence."[5] The claim for methane on Mars, and in particular the claim that the abundance of methane in the Martian atmosphere can vary with location and rapidly with time, should, according to Zahnle, "be regarded as an extraordinary claim."[6] On that point, virtually every planetary scientist on Earth would agree with Zahnle. The question, then, reduces to this: has the Curiosity team provided extraordinary evidence?

For methane to appear in the Martian atmosphere as variable in both time and location, several requirements must be met. First, one or more local sources of methane must exist. Second, those methane sources must be capable of outgassing methane into the atmosphere in copious quantities on a timescale of weeks to months. Third, one or more sinks of methane must exist that can remove methane from the atmosphere on a similar timescale so that methane levels do not build up in the atmosphere. Do such sources and sinks exist?

Methane has been studied rigorously in what we can call conventional atmospheric photochemistry models. That is, how will methane gas behave in an atmosphere when surrounded by other gases such as those that are present in the atmosphere of Earth (composed of mostly nitrogen, oxygen, argon, carbon dioxide, and water) or Mars (composed mostly of carbon dioxide, nitrogen, argon, oxygen, and carbon monoxide), at the pressures and temperatures appropriate for these atmospheres and when exposed to the amounts and kinds of sunlight expected for such environments? The answer: in both atmospheres, methane will react with the other molecules very slowly but very dependably until it is destroyed.

In about 350 years, any methane injected into the Martian atmosphere should have reacted with other gases and been converted into other molecular species. Over a time span of much less than several centuries, Martian weather will disperse any methane emitted in any single location globally. At any particular time, the entire atmosphere should show approximately the same level of methane.

The process of destroying methane molecules is not instantaneous. It would occur continuously but would almost certainly not produce measurable changes on a timescale of days, weeks, months, or even years. Such changes would only be detectable as individual events accumulate over decades. Yet the changes detected on Mars by the Curiosity experiment in early 2014 occurred in a matter of weeks, thousands of times faster than "over decades." Clearly, Zahnle asserts, the process of change, in which the observed levels of methane change substantively on a timescale of one to two months, cannot be attributed to conventional atmospheric photochemistry. Something else must be at work.

If Martian methane is not destroyed slowly by conventionally understood photochemical processes, perhaps it is destroyed rapidly
by reacting with oxygen. Chemists would agree that oxidation reactions* could quickly remove the methane; of course, such reactions
require a source of oxygen and the atmosphere of Mars is deficient
in oxygen. Unlike the atmosphere of Earth, in which oxygen is generated continuously by photosynthetic reactions in plants, Mars has no
plant life (that we know of) and almost no free oxygen (0.13 percent)
in its atmosphere; on Mars, the only viable source of oxygen release
into the atmosphere is the photolysis of water, that is, the breaking
apart by sunlight of H_2O into its constituent atoms of hydrogen and
oxygen. This process would generate, in addition to oxygen atoms,
free hydrogen atoms. The free hydrogen atoms, because they are so
lightweight, would bubble up to the top of the Martian atmosphere
and escape to space. If this process is ongoing at a substantive level,
we should be able to infer this activity from the rate at which hydrogen atoms escape from Mars into interplanetary space.

In fact, planetary scientists have done that. The rate at which free
hydrogen atoms escape from the atmosphere of Mars has been measured and is known.[7] Assuming all the hydrogen atoms that scientists detect escaping from Mars come from the breakup of water
molecules, then the number of escaping hydrogen atoms tells us,
also, the production rate of free oxygen atoms that become available
to oxidize methane molecules. What, then, is the answer? The known
production rate of oxygen atoms, based on the escape rate of hydrogen atoms, is ten times too small to account for the rate at which the
methane is destroyed, assuming methane is regularly depleted from
30 parts per billion to zero parts per billion in four months, which
is the level of change and the timescale consistent with the claimed
changes observed by multiple observers. If the 30 parts per billion
measurements are wrong, but the 7 parts per billion measurements
by Curiosity are right, the oxygen production rate is still about three
times too small. Overall, the evidence strongly suggests that the oxidation of methane molecules by free oxygen is not a viable sink

---

*The terrestrial process by which oxygen reacts with iron to form iron oxide (or rust) is an
example of an oxidation reaction.

for removing methane from the atmosphere unless almost all of the claimed detections of methane are wrong.

Possibly, as suggested (without any evidence) by Geminale, Formisano, and Sindoni in presenting their Mars Express results, the oxygen-rich dust on the surface could provide the reactive oxygen that might destroy methane molecules near the surface. These same authors also suggested that fast-moving electrons generated by electric fields in swirling dust devils during Martian dust storms might provide a quick-acting mechanism to destroy Martian methane. Such ideas are imaginative and untested, but at least are marginally plausible and worthy of further investigation.

In 2015, James Holmes, Stephen Lewis, and Manish Patel proposed the possibility that zeolites could be fast-acting methane absorbers.[8] Zeolites are porous aluminosilicate minerals that act like sieves, sorting by size molecules with which they come into contact. In industrial usage, zeolites can be used to purify water and separate molecules of different sizes. On Mars, zeolites that could trap methane could be present in both the atmosphere and the loose, uncompacted layer of dust and rock fragments (the regolith) that covers most of the surface of the planet. Martian dust devils or dust storms could put zeolites in contact with atmospheric methane, whence they could quickly remove the methane from the atmosphere.

Zahnle asks whether we should believe the measurements of methane. He notes that the problems with the claims for methane detections are many. For example, with the exception of the Curiosity results, the claims for detections of methane are based on heavily modeled spectra. For the observations made from Earth, the observers must remove the effects of Earth's atmosphere on the spectra. Yet, those models are imperfect. The light from Mars must pass through Earth's atmosphere, which contains hundreds to thousands of times more methane than does the atmosphere of Mars. Even though the Martian methane lines are Doppler shifted slightly from their terrestrial counterparts, almost none of the Martian lines (if any are real) have been seen, and those few spectral wiggles that have been identified as the spectral signature of methane may nevertheless still be random noise, not signal. Additional problems lie with the corrections made in the modeled spectra based on the temperature-sensitive water lines in Earth's atmosphere, and those terrestrial water signals, which overlap the Martian methane lines, are "far stronger than the putative Martian signal."

On the other hand, the measurements made by the Curiosity rover from the surface of Mars are not susceptible to any errors introduced by these enormous modeling problems. Notably, the Curiosity team reported in 2013 that they did not detect any methane on Mars, and their nondetections were at such a low sensitivity level that they suggested that all of the previous detections of methane were suspect. Alternatively, one could understand their original nondetections to mean that Mars episodically releases methane and that one has to get lucky to detect it by observing at just the right time and being in just the right place. By this logic, the Curiosity rover was in exactly the right place at exactly the right time in early 2014, and so detected a burp of methane, albeit a modest burp, during a brief interval of time. Again, if Curiosity has detected Martian methane, the Curiosity results seemed to provide some support to the possibility that Mars experiences small methane eruptions. That's a big "if."

"I am convinced that they really are seeing methane," Zahnle said, in reference to the Mars Curiosity rover report of 2015, "But I'm thinking that it has to be coming from the rover." That is, Curiosity brought the methane with it from Earth and is detecting terrestrial methane, not Martian methane. Even Chris McKay, another NASA Ames scientist and one of the co-authors of the January 2015 paper on the Curiosity results, agrees that Zahnle's concerns are valid: "I think the possibility of a methane source aboard should still be considered until completely ruled out."[9] Of course, even Webster and his TLS team acknowledge bringing methane from Earth to Mars aboard Curiosity, but they also believe they were able to completely remove its effects when performing their analysis of their data. They also argue that they have successfully pumped most of the methane out of the Herriott cell so that any contribution to the measured methane signal from terrestrial methane is negligible; furthermore, they consider and reject as a source of methane chemical reactions with the surface coatings of the chamber and the internal mirrors. Continuing through their thorough analysis, they evaluate and dismiss the possibility that the methane they detected was generated by the degradation of the rover's wheels or by the rover crushing Martian rocks. After all considerations, they conclude, "Our measurements spanning a full Mars year indicate that trace quantities of methane are being generated by Mars by more than one mechanism or a combination of proposed mechanisms—including methanogenesis either today or released from past reservoirs, or both." That is an extraordinary claim.

As for Zahnle's critique of the Curiosity result, most of the detected methane, Zahnle asserts, "is in the antechamber to the sample cell and comes from several sources, known and unknown, in the rover itself."[10] Zahnle examines the data from the rover results released in 2015 and notes:

> At first, no Martian methane was seen, both while the rover was awash in stowaway Florida air and then after the rover was evacuated. But later, as methane slowly built up again inside the rover, methane appeared in five of six samples of Mars's air at levels on order of 7 parts per billion by volume (ppbv).... After the fifth sighting, TLS/SAM performed the first of two higher-sensitivity enrichment experiments, only to find the methane nearly gone.... William of Ockham* would warn us to be wary of peekaboo methane when a known source—the rover—is so nearby. Because the concentration of methane inside the rover is approximately 1,000 times as high as that in the Martian air, it would not take that much. But it would be a mistake to be too confident that the methane was never there.[11]

Zahnle also points out that the SAM team performed two experiments designed to make the Martian methane easier to detect—and both times they detected no methane. These "enrichment" experiments (on sols 573 and 684) were done by scrubbing the air in the Herriott chamber of carbon dioxide. Since 96 percent of the Martian atmosphere is carbon dioxide, by removing virtually all of that gaseous species, any remaining molecular species should be 10–20

---

*William of Ockham (1280–1349 CE) was a critic of knowledge, best remembered for having said that simple explanations are preferred over more complicated ones; therefore, a theory should be no more complicated than absolutely necessary.

times easier to detect. On both sol 573 and 684, the methane level
was below 1 part per billion. Zahnle concludes that the TLS team
has not presented extraordinary evidence to support their extraordinary claim.

nonbiological sources of methane

What if the methane levels detected on Mars are correct and the
origin of that methane is Mars itself? If we assume that present-day
Mars does have extremely low levels of methane in its atmosphere
that is variable in both time and space, we should reasonably follow that assumption with another question. Is biological activity
the only possible source of that methane? The answer is no. Scientists have identified several other methane-production pathways as
plausible.

Serpentinization. If subsurface water reacts with basaltic rocks (e.g.,
with the mineral olivine* or pyroxene**), the basalt is turned into the
mineral serpentine.*** This process of serpentinization liberates hydrogen atoms, which, in the Martian environment, would combine
with carbon (available abundantly in the atmosphere in the form of
carbon dioxide molecules) to create methane. Planetary geologists
agree that this process almost certainly actively injected significant
amounts of methane into the Martian atmosphere when Mars was
young. The high levels of the greenhouse gas methane may have
helped raise the temperature of the atmosphere of a young Mars to
a level high enough to allow the presence of oceans and a global

---

*Olivine is one of the most common minerals on Earth and is likely similarly abundant on
Mars. An olivine mineral, identified chemically as $(Mg,Fe)_2SiO_4$, includes one silicon atom,
four oxygen atoms, and most commonly either two magnesium atoms (Mg_2SiO_4; forsterite)
or a mixture of Si and Mg atoms. An olivine mineral can also contain calcium ($CaMgSiO_4$;
monticellite) or iron ($CaFeSiO_4$; kirschsteinite).

**A pyroxene mineral includes six oxygen atoms and usually two silicon atoms, though alu-
minum could substitute for the silicon atoms. A pyroxene mineral (identified chemically as
$(NaCa)(Mg,Fe,Al)(Al,Si)_2O_6$) also includes two other elements, one of them typically sodium
or calcium and the other either magnesium, iron, or aluminum.

***A serpentine mineral contains magnesium, silicon, oxygen, and hydrogen: $Mg_3Si_2O_5(OH)_4$.

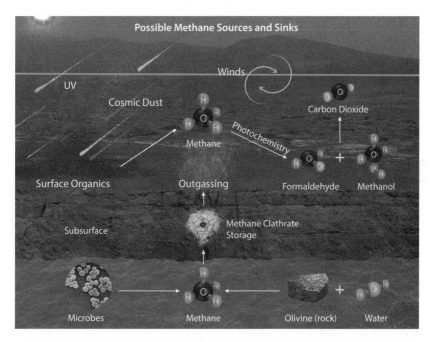

Figure 15.2. Illustration of several possible ways that methane might be added to and removed from the Martian atmosphere. Subsurface sources include methanogenic microbes, serpentinization (water reacting with olivine minerals), and ancient methane escaping from clathrates. Surface sources include ultraviolet light reacting with carbon-rich dust that rains down onto the surface. Mechanisms that might destroy methane include surface chemical reactions that utilize oxygen-rich molecules to process methane into carbon dioxide and swirling dust devils that generate fast-moving electrons that act to destroy the methane. See Plate 10. Image courtesy of NASA's Goddard Space Flight Center/Brian Monroe.

water cycle. Much of that methane, produced by a young Mars, may continue to be sequestered in permafrost, from which it could be steadily released.[12] One estimate for the maximum amount of methane that could be released, annually, by the process of serpentinization is 2,000 times greater than the 100 tons per year necessary to generate 10 parts per billion of methane.[13]

Photolysis of water vapor. Akiva Bar-Nun and Vasili Dimitrov, of Tel-Aviv University, showed that the photolysis of water vapor in the presence of carbon monoxide can lead to the production of methane and other hydrocarbons.[14] The first step in the process occurs when a water molecule absorbs an ultraviolet photon and is separated into an OH radical* and a hydrogen atom. The OH radical reacts with CO to form a carbon dioxide molecule (CO_2) and release a hydrogen atom. Some of the freed hydrogen atoms will react with other CO molecules, with methane as the final product. The production of methane by this process is incontrovertible. Whether this process actually occurs on Mars is the important and relevant question. Bar-Nun and Dimitrov argued convincingly that it can and should occur under actual Martian atmospheric conditions, and can produce methane at levels thousands of times greater than the levels purportedly observed by those who claim to have detected methane in the Martian atmosphere.

Meteoritic sources. Certain meteorites contain up to several percent carbon in carbon-containing compounds. Mars, being the closest planet to the asteroid belt, which is the part of the solar system in between the orbits of Mars and Jupiter, experiences a great deal of bombardment by meteoritic material, very little of which burns up as it descends through the thin atmosphere of Mars. Hence, the surface of Mars must be covered with a thin layer of dust composed mostly of meteoritic debris. When exposed to ultraviolet light from the Sun, as it will be on the Martian surface, and under light conditions designed to mimic those found on Mars, researchers found that carbon-rich meteoritic material from the meteorite known as Murchison released methane very efficiently. The amount of methane released by Murchison increased with increasing temperature. Frank Keppler, of the Department of Atmospheric Chemistry at the Max Planck Institute for Chemistry in Mainz, Germany, and his colleagues estimated that the amount of meteoritic material that falls unaltered to the surface of Mars and is then exposed to ultraviolet light, thereby becoming a source for methane production, is more than sufficient to account for an equilibrium level of about 10 parts per billion of methane in the atmosphere of Mars.[15]

Methane clathrates. Clathrates form when liquid water solidifies under conditions of both low temperature and high pressure. When this happens, cage-like cavities form within the ice. These

---

*A radical is an element or compound with an unpaired electron that, as a result, is very reactive.

cages become reservoirs that can trap gases like methane. On Mars, methane-rich clathrates could have formed early in the history of Mars, when the Martian atmosphere was hypothesized to have been richer in methane. Such clathrates are thought to exist on Mars, beneath the surface and at the polar caps, and to have played an important role in the deep past for the climatic evolution of Mars. One plausible source of Martian methane, therefore, may be the slow escape of methane gases from subsurface reservoirs of clathrates. Given the small amounts of methane supposedly detected in the present-day Martian atmosphere, the spatial extent of the necessary methane clathrate deposits from which methane gas would be seeping would be quite small.[16]

Does Mars have methane? And is that methane proof of life on Mars? In the end, we have conflicting evidence for the presence of methane in the modern atmosphere of Mars. If it is present at all, the evidence, controversial as it is, suggests it appears as the result of episodic eruptions of methane into the atmosphere. Then, within weeks to months, the methane disappears.

A steady-state low level of methane in the Martian atmosphere can be generated through a number of processes, none of which require life on Mars now or in the past. But life could also produce this level of methane. The rapid disappearance of Martian methane, however, is much more difficult to explain than the continuous presence of low levels of methane, no matter the production source of the methane. Without a viable mechanism to rapidly remove methane from the Martian atmosphere, all claims for Martian methane appear to lack enough supporting, extraordinary evidence.

Curiosity itself has now been on Mars for long enough to begin to sort through the possible methane-generating possibilities. If the methane is produced by subsurface methanogens, then methane production is likely to follow a Martian seasonal cycle. Thus, since the first methane surge occurred in autumn in the southern hemisphere of Mars (December 2013 through January 2014), then it likely should have occurred again during the same part of that same Martian season, i.e., Martian southern autumn during late 2015 into early 2016. According to a NASA press release of May 11, 2016, "the mission checked carefully for a repeat of [the methane surge seen in 2013–2014], but concentrations stayed at lower background levels" in 2015–2016. Team leader Chris Webster said, "Doing a second year told us right away that the spike was not a seasonal effect. It's apparently an episodic event that we may or may not ever see again."[17] These new measurements suggest that the methane burp of 2013–2014 was a random outburst or perhaps contamination, and they reduce but do not eliminate the likelihood that it is associated with life on Mars.

16

---

chasing martians

Could life exist on Mars today?

Certain conditions must hold true in order for chemically based life to develop and take root. The environment almost certainly must have liquid water and a source of energy. Mars has ancient, dried-up river valleys, deltas, and lakelike formations, and the Sun shines brightly on Mars. Check.

The atmosphere and soil must supply a handful of bio-essential elements: carbon, oxygen, nitrogen, hydrogen, phosphorous, and sulfur. Mars has an abundant supply of all of these elements. Check.

The environment must maintain these conditions for a long enough time for life to develop. Tantalizing evidence, summarized in 2016 by Ray Arvidson, the James S. McDonnell Distinguished University Professor of Earth and Planetary Sciences at Washington University in Saint Louis, reveals multiple time periods when Mars was warm enough and wet enough for long enough to support life.[1] The Opportunity rover explored the ancient (4-billion-year-old) remnants of the 14-mile (22-kilometer) wide Endeavor Crater and found, he writes, that "the formation of this crater generated an extensive and long-lived hydrothermal system . . . that would have produced a relatively habitable subsurface environment, at least in terms of sustained availability of water." Opportunity also explored the 3–4-billion-year-old Burns Formation Outcrops, which show multiple signatures (e.g., ripple patterns and sedimentary rocks) that are evidence for the long-term presence of surface and subsurface water. Arvidson explains that "the wet surface environments, if on Earth, would have been habitable, if only for microorganisms adapted to acidity and long periods of aridity. . . . The subsurface groundwater . . . would have had more clement conditions and if on Earth surely would have been habitable." Arvidson explains that the Spirit rover explored a region of rolling hills surrounding an eroded, volcanic region known as Home Plate, a region dated to be no older than 3.7 billion years. "On Earth all of the aqueous environments around Home Plate would be habitable and may have been for Mars." The Curiosity rover explored a region of deposits inside the 93-mile (150-kilometer) wide Gale Crater, including the region of sedimentary strata around Mount Sharp, the mountain at the center of this large crater where some of the rocks are as young as 2.9 billion years. Curiosity's measurements reveal that Gale Crater was once an enormous lake where mudstones now overlay sandstones. The Gale Crater region deposits formed at a time with "significant surface runoff and associated erosion and transport of sediment." The Curiosity data "point to environments conducive to preservation of organic molecules and ones that were habitable, at least in terms of sustained presence of water." Given that "all of these environments would likely have been habitable, at least in terms of the availability of water," we know that many regions on and under the surface of Mars were habitable, if not inhabited, for at least the first 1.5 billion years of Martian history. Check.

No definitive evidence for the presence of past or present life on Mars has yet been discovered. We are still waiting for that proof or for clear and strong evidence that says Mars is and has always been barren. But without any doubt, Mars was once hospitable to life; it was a planet on which Earthlike forms of life might have flourished. The combination of the rover discoveries with the possible (though not convincing) evidence of ancient, or even present, Martian life as seen in ALH 84001, and with the possible (though not convincing) measurements of fluctuating atmospheric methane abundances that might be at levels above that which could be produced without biological activity demonstrate that we absolutely cannot definitively deny that life might have existed and might still exist on Mars. We are still chasing Martians.

Indeed, the idea that Mars once had life is not far-fetched, and if Mars once hosted living things, those living things could have adapted to the changing Martian climate and found a way to survive all the way until the present day. We can very reasonably conclude that the idea that Mars might host colonies of subsurface microscopic life-forms is plausible. Therefore, the most important questions about Mars today are, Did the known habitable regions have the "bioavailable compounds" and sources of energy necessary to sustain life? and, Does any form of life exist on Mars now?

If we have Martian neighbors, they have been teasing and testing us for 400 years. Once the age of telescopes arrived, we discovered what some astronomers thought were clues to the presence of Martians in the form of the canals stretching across the surface. Yes, the evidence was indirect. No, we didn't see any Martians, but the apparent evidence was hard to miss and even harder to ignore, at least to those who actually thought the lines on the surface of Mars were canals and believed in Martians. Of course, those surface markings were never as long or as straight as Giovanni Schiaparelli and Percival Lowell thought they were, and Mars never had any artificial canals. No evidence of life.

In the twentieth century, astronomers' certainty as to exactly what life-forms they might find on Mars changed. They developed a strong consensus that Lowell was wrong and that Mars did not host intelligent, canal-building engineers, but perhaps Lowell and others were right about Mars having vast forests or large swaths of quick-growing vegetation. Some astronomers then found alleged evidence for vegetation on Mars via the presence of chlorophyll, or at least via the presumed effect of chlorophyll on light reflected from Mars. Disappointment returned again when other astronomers showed that Mars was never green and that the reflected light from Mars was inconsistent with the presence of chlorophyll. No chlorophyll; thus, no evidence of life.

A few years after astronomers lost their enthusiasm for chlorophyll, they claimed decisive spectroscopic evidence first for lichens and then for algae, both of which are more primitive life-forms than photosynthesizing plants. But the spectroscopic evidence was misunderstood and misinterpreted. No lichens. No algae. No evidence of life.

Mars today is a tough environment for life. It is cold, with liquid water on the surface only rarely, if ever. The atmosphere is thin and provides little protection from high-energy photons and particles. If any Martian life-forms that once thrived have survived, they must be hardy but also hard to spot. Modern Martians might all be microscopic; in addition, because the Martian atmosphere cannot protect them from the Sun's deadly ultraviolet rays, they might need to hide under rocks or burrow deep underground in order to secure the warmth and water necessary for their survival and to protect themselves from dangerous radiation from space.

Even subsurface, microscopic Martians, however, need to inhale and exhale. The chemical waste products from their respiratory activities should build up in the soil and eventually, slowly but surely, seep into the atmosphere. Even at the level of one biological tracer molecule out of every billion other nonbiological molecules in the atmosphere of Mars, we should be able to sniff out the presence of those biologically produced molecules. Even at the detectability threshold of only a few parts in a trillion, we have developed the ability to ferret out the evidence of the effluence of microscopic Martians with the high-tech sniffers and probes we have sent and continue to send to Mars on our landers and rovers.

The atmospheric methane gas astronomers thought they had detected offered evidence to anyone who understood the details of that kind of data that Martian life-forms were affecting the chemical makeup of their atmosphere by inhaling and exhaling. Using cameras on our rockets to and landers on Mars, we have not seen any macroscopic creatures, Carl Sagan's hypothesized Martian macrobes, wandering around on the surface of Mars, but a high level of atmospheric methane appears to demand microscopic biological activity as the source. If Mars does have substantial and changing levels of atmospheric methane, the Martians may have been found.

As it turns out, the first and second and third "discoveries" of methane were likely all wrong. Almost certainly, none of these were actual detections of methane. But, like a phoenix rising from the ashes, claims for definitive discoveries of methane in the atmosphere on Mars have returned again and again. Did Krasnopolsky and his collaborators detect low levels of methane in 1988? In 2003? In 2006? Maybe. Did Mumma and his collaborators detect low levels of methane in 2003 and 2006 and 2009? Maybe. Did the Mars Express team detect low levels of methane in 2004? Maybe.

Are the most recent measurements from the Curiosity rover that purportedly are evidence for a slightly enhanced level of methane in the Martian atmosphere over a brief, two-month period indisputably detections of methane? Yes. Are the arguments identifying methanogenic, subsurface bacteria as the source of this methane indisputable? No. Can this evidence be ignored? No way.

Is the actual steady-state level of methane in the Martian atmosphere, that being the amount found in between methane hiccups (if those hiccups are real), in the range of 0.1 to 1.0 parts per billion, as detected by Curiosity? Probably. Is this amount of atmospheric methane consistent with the continuous input of organics from meteorites, the inorganic production of methane through serpentinization, and with the several-centuries-long photochemical lifetime of methane in the Martian atmosphere? Yes. The Martians have not been found, and maybe they are not there to be found. We simply don't know, yet.

Perhaps Mars used to have life. Some scientists are convinced that our Martian neighbors sent us evidence, in the form of a meteorite rocketed off the surface of Mars, that says, "we slept here." After orbiting the Sun for about seventeen million years, that meteorite landed on Earth, in Antarctica. Thirteen thousand years later, we found it and sent it to one of our geochemical laboratories already equipped to study moon rocks and to find tiny traces of rare elements in meteorites. In this Martian meteorite, our Martian neighbors did not simply send us indirect evidence, from which we could infer their presence. No, they offered certainty by sending to us what might be self-portraits in the form of fossils. Incredibly, we now have fossil evidence of ancient life on Mars, provided you believe these are actually fossils. After two decades of passionate debate based entirely on new scientific evidence that continues to emerge, all of the fossil evidence has now been called into question. Almost all scientists think the rod-shaped fossils of "bacteria" are merely interestingly shaped, non-biological minerals. Is all the evidence for life found in the meteorite known as ALH 84001 absolutely, indisputably wrong? No. Most of these claims have withered under intense scientific scrutiny, but a very small possibility exists that some of the evidence, in particular the presence of tiny magnetite grains, could be evidence of life. The chances that the clues in ALH 84001 point to ancient life on Mars are slim, but they are not nonexistent. Possibly ancient Martians have been found.

The Outer Space Treaty, adopted by the United Nations in 1967, includes the following principles, among others: "The exploration and use of outer space, including the moon and other celestial bodies, shall be carried out for the benefit and in the interests of all countries" and states shall "conduct exploration of them so as to avoid their harmful contamination."[2] We might ask, today, whether we know enough about Mars to be certain that current plans to send humans to Mars are consistent with these principles in the Outer Space Treaty. If the possibility exists that any life exists on Mars that is native to Mars, might it be in the interests of all countries to avoid further contamination of Mars until we can answer, with great certainty, the question, Did or does life exist on Mars?

Today, NASA's Curiosity rover continues to probe the atmosphere of Mars, testing for Martian methane and identifying locations on Mars where life might once have flourished and might still exist. New robotic spacecraft are under construction or on drawing boards that could explore these locales and look for signs of life. Telescopes continue to peer down from Martian orbit, looking for more clues. Tomorrow, next week, next year, or next decade we may finally discover that we are not alone in the universe and that Mars, our close planetary neighbor, has secrets it can no longer keep.

Chasing Martians is very much a We-the-People activity. We the People believe in Martians. Sometimes We the People are the Martians, as was astronaut Mark Watney in the 2015 movie *The Martian* and as are the First Hundred and their descendants in the *Red Mars, Green Mars, Blue Mars* trilogy written in the 1990s by Kim Stanley Robinson. Sometimes we are descended from ancient alien Martians, as we all are in the 2000 movie *Mission to Mars*. Sometimes we find little Martian plants, as did the astronauts in the 2016 National Geographic mini-series *Mars*. Other times, when we first venture to Mars, we experience a dangerous encounter with alien Martians, as do the astronauts in Greg Bear's short story *Martian Ricorso* (1976).

We the People want and expect scientists to find life on Mars. Some of us think we are looking to find a pathway toward our future, our destiny, because the existence of life on Mars, now or in the past, would make the possibility that humans will colonize and survive on Mars seem more likely. These are good motivators for further study of Mars. They also provide good reasons for humanity to voluntarily delay our collective ambitions to colonize Mars and give the scientific community enough time to learn, to a high degree of certainty, whether Mars is and always has been sterile. If Mars is barren, we would no longer have any reason to restrain ourselves in colonizing and attempting to terraform Mars; however, if Mars hosts life, then humanity's future relationship with Mars becomes more complicated.

In the purest sense, we have already contaminated Mars. The Committee on Space Research of the International Council on Science (COSPAR),[3] which was founded in 1958, established rules that required spacecraft headed to the surface of Mars to be sterilized. After the Viking missions revealed no apparent evidence of life and uncovered a Martian world that appeared inhospitable to life, COSPAR changed the rules for Mars. Today, Martian landers must achieve a high level of cleanliness, but they need not be sterilized. Yes, modern Martian landers are assembled in "clean" rooms; yes, Martian landers are likely nearly sterile. But they are not sterile. Some terrestrial bacteria have already hitched rides to Mars on the exposed, outer surfaces of spacecraft where most of them will die from exposure to solar ultraviolet radiation before arriving at Mars. Others, however, will reach Mars as stowaways inside spacecraft where they might survive for thousands of years. NASA's Chris McKay estimated, in 2007, that the Exploration rovers, Spirit and Opportunity, each were host to as many as 100,000 "microscopic Earthlings."[4] McKay pointed out that the first Martian colonists from Earth arrived on July 4, 1997, when NASA's Pathfinder rover touched down on Mars. McKay also argued that these terrestrial visitors are "incapable of growing and spreading. They cannot grow because there is no liquid water, and they cannot spread, because once released into the environment, they are rapidly killed by the Martian ultraviolet light." If we are lucky, McKay is right. Even if McKay is wrong, we can still safely assume that virtually the entire Martian surface remains uncontaminated by terrestrial life-forms. Meanwhile, despite this minor level of contamination, we should allow the scientists in the laboratories studying meteorites, the astronomers on mountaintops collecting data with their telescopes, and the engineers in machine shops who are building new rovers and detectors to strap into rockets and send to Mars, to continue their work and find answers before we (further and irreparably) contaminate Mars.

Mars may be our destiny. But not yet. We can and should continue to study and explore Mars, but for now we do not need to contaminate Mars any more than necessary.

If any chance remains that Earth is not the only world in our solar system with active biology, that another world might have developed life independently of Earth, if life-forms of Mars may once have flourished but now lie low, we need to preserve our chances of confirming this possibility before it is too late.

By chasing Martians, we have already learned a great deal, even if we don't yet know whether Martians exist. Percival Lowell, motivated by his study of the supposed Martian canals, built Lowell Observatory and that is where, in 1930, Clyde Tombaugh discovered Pluto. Infrared astronomy developed rapidly in the 1950s, at least in part thanks to the interests of the pioneers Gerard Kuiper and William Sinton, who wanted to push infrared astronomy forward so that they could study and better understand the colors of Mars. Dozens of scientists, inspired by the possibility of an extreme form of life in ALH 84001, have discovered a dizzying array of creatures known as extremophiles that live in extreme environments here on Earth, including thermophiles that live in very high-temperature environments, halophiles that live where salt concentrations are excessively high, and acidophiles that thrive in high-acid locations. Planetary scientists, motivated to understand whether life could exist in the Martian environment, have identified Europa, Enceladus, and

Titan as places in the solar system where life might exist independent of sunlight and solar heating. Scientists don't always discover what they are looking for when they design their experiments, but once an experiment is under way they almost always discover things worth knowing. Our pursuit of Martians is a century-long example of curiosity-driven science leading to important discoveries.

On the critical question of whether life exists on Mars, the jury is still out. We have to answer the questions, "Does Mars now or did Mars ever have life?" and "What is Martian life like?" before we destroy the evidence and before we decide whether colonizing and terraforming Mars is an activity we humans should undertake. Until and unless the preponderance of the evidence we are able to collect, using remote and robotic tools, indicates Mars is sterile, we should heed Carl Sagan's admonition: "If there is life on Mars, . . . Mars then belongs to the Martians, even if the Martians are only Microbes."[5]

acknowledgments

I am grateful to Dr. Michael Lipschutz, emeritus professor of inorganic and cosmochemistry at Purdue University, who carefully read this manuscript several times and, in doing so, tried to teach me some basic chemistry I learned once but forgot long ago, and who also provided me with valuable pointers into the history of meteorites. I also thank my friend, bike-riding companion, and colleague, Bob O'Dell, astrophysicist and former chief scientist for the Hubble Space Telescope, for his very careful review of these pages and for prodding me to select some of my words more carefully: absorption lines and absorption bands are different, as are photographs and images. Also, thank you to my former student Jacob Graham, who read an early draft of this manuscript as part of a "directed readings" course in astronomy and who tried to make sure my writing was readable by nonexperts like himself. If any sections of this book are not, don't blame Jacob. I am fortunate to have two authors as parents. They inspire me, repeatedly, to do something as foolish as write a book, continually send me news clippings that might be of value in my work, find endless grammar mistakes in my writing, and remind me that many of my sentences—like this one—are too long and convoluted and should be shortened. As always, I offer my love and gratitude to my wife, Carie Lee, who puts up with me and has the tough chore, on a daily basis, of keeping me from spending too much time thinking about heavenly bodies.

life on mars astronomers and scientists

Edward Anders: meteoriticist at University of Chicago who in 1964 debunked the claim that the Orgueil meteorite, which fell in France in 1864, contained organic material native to the meteorite.

Eugène Michael Antoniadi: Turkish-born astronomer who, as leader of the Mars section of the British Astronomical Society in the 1890s and 1900s, initially confirmed the existence of canals and areas with grassy vegetation on Mars but later became leader of the anti-canal movement among professional astronomers.

E. E. Barnard: American astronomer who, from his own observations of Mars in the 1890s and in 1910 and his inspection of Lampland's 1905 photographs of Mars, was unable to identify anything resembling a canal on Mars.

Father Daniello Bartoli: Jesuit priest who, on Christmas Eve 1644, detected two patches on the "lower part" of Mars.

Wilhelm Wolff Beer: German banker who, working with Johann Heinrich von Mädler, used observations of Mars made between 1831 and 1839 to create the first global map of Mars.

Klaus Biemann: MIT scientist and team leader of gas chromatograph and mass spectrometer experiment on the Viking landers that found no organic molecules in the Martian soil.

C. E. Burton: Irishman who in 1882 reported the first confirmation of the doubling of the canals of Mars, first reported by Schiaparelli.

William Wallace Campbell: American astronomer who in 1894 used spectroscopic observations of Mars and the Moon to show that the water vapor seen in spectra of Mars was actually in Earth's atmosphere; later, as director of Lick Observatory, in California, he led an observing campaign in 1908 that proved Mars had no detectable water vapor in its atmosphere.

Giovanni Domenico Cassini (also Jean Dominique Cassini): papal astronomer and the first director of the Paris Observatory who in 1666 used observations of two dark patches on Mars to determine that the rotation period of Mars was 24h 40m.

Robert N. Clayton: American meteoriticist who in 1993 confirmed the Martian origin of ALH 84001 by analyzing the isotopes of oxygen in the meteorite.

William Weber Coblentz: American physicist who in the 1920s used infrared light reflected from the surface of Mars to make the first accurate measurements of the surface of Mars and used these temperatures as an argument in favor of vegetation that can grow in dry, cold environments.

Norman Colthup: American chemist who in 1961 claimed to have proven that the Sinton bands were the result of fermentation processes by organic matter on Mars.

Pierre and Janine Connes: French chemists and astronomers who in 1966, with Lewis Kaplan, reported at two conferences the discovery of methane in the atmosphere of Mars that they and the press reported as evidence of life on Mars; though they never retracted their work, they also never published their results.

C. H. and Edith L. R. Corliss: members of a National Bureau of Standards observing team, along with C. C. and Harriet K. Kiess, who in 1956 made observations of Mars from near the summit of Mauna Loa in Hawai'i that proved that the amount of water in the atmosphere of Mars was below the threshold of detectability at that time.

William Rutter Dawes: English physician, clergyman, and astronomer whose work in the mid-1860s revealed that the red color of Mars was due to something on the surface rather than something in the atmosphere.

Camille Flammarion: French astronomer and popularizer of astronomy who in 1862 first published *Plurality of Inhabited Worlds*, in which he articulated reasons why other planets were populated and pointed out that the many similarities between Earth and Mars led very naturally to the conclusion that Mars is inhabited by intelligent beings.

Francesco Fontana: Neapolitan lawyer, optician, and amateur astronomer who in 1636 discovered a dark spot in the middle of the disk of Mars that was almost certainly a result of the poor lenses in his primitive telescope.

Bernard Le Bovier de Fontenelle: French author who published *Conversations on the Plurality of Worlds* in 1686.

S. Fonti: Italian astronomer who, with G. A. Marzo, used the Thermal Emission Spectrometer measurements made in 2000–2002 by the Mars Global Surveyor spacecraft to report, in 2010, having detected methane in the Martian atmosphere; in 2015, he retracted that claim.

Vittorio Formisano: Italian astronomer and leader of Planetary Fourier Spectrometer for the Mars Express mission who in 2004 reported detection of Martian atmospheric methane.

Everett Gibson, Jr.: NASA planetary scientist and astrobiologist; *see* David McKay.

Nathaniel Green: English artist who in 1877 published a detailed map of Mars.

Father Francesco Grimaldi: a Jesuit who, working with Father Giambattista Riccioli at the Collegio Romano in Rome, reported observing patches on Mars on multiple nights in 1651, 1653, 1655, and 1657.

William Haseltine: American chemist who in 1965, along with James Shirk and George Pimentel, proved that the Sinton bands were caused by water in Earth's atmosphere rather than by algae on Mars.

William Herschel: German-English astronomer who in the late eighteenth century established that Mars has Earthlike seasons, with northern winter occurring at the same time as southern summer, when he discovered that the sizes of the bright northern and southern polar spots waxed and waned antisynchronously; also discovered the tilt of the rotation axis of Mars with respect to the orbital plane of Mars around the Sun, measured the rotation period to be 24h 39m 21.67s, and proved that Mars has an atmosphere.

Norman Horowitz: biologist at Caltech and team leader of Pyrolytic-Release experiment on Viking landers that reported no evidence of life on Mars.

William Huggins: astronomer and professor of chemistry at Kings College London, who in the 1860s applied the newly invented technique of astronomical spectroscopy to his observations of Mars that produced evidence for the presence of water vapor in the atmosphere of Mars.

Christiaan Huygens: Dutch astronomer who in 1659 identified a large, broad, dark "V"-shaped area on Mars and, by watching that patch, ascertained that Mars rotates in about 24 hours.

Jules Janssen: French astronomer who in 1867 used the technique of spectroscopy to conclude that the Martian atmosphere contains considerable quantities of water vapor.

Lewis Kaplan: American astronomer who in 1963, with coworkers Guido Münch and Hyron Spinrad, used the 100-inch diameter telescope on Mount Wilson in California to make the first definitive detection of water vapor in the atmosphere of Mars; *also see* Pierre and Janine Connes.

C. C. and Harriet K. Kiess: *see* C. H. and Edith L. R. Corliss.

Harold Klein: director of Life Sciences at NASA's Ames Research Center and leader of NASA's biology investigation team for the Viking program who concluded that the Viking experiments revealed no evidence for life on Mars.

Vladimir Krasnopolsky: astronomer who, with Michael Mumma, in 1997 reported having detected the possible presence of methane on Mars with data collected with a telescope in Arizona in 1988; and in 2004 reported having detected methane on Mars in 1999 observations using a telescope in Hawai'i and, having calculated that methane could not survive for more than a few hundred years in the Martian atmosphere, that the methane was evidence of active biology on Mars; and in 2012 again reported detecting Martian methane in 2009.

Gerard P. Kuiper: Dutch-American astronomer who obtained the first-ever color photograph of Mars in 1948 and over the next decade presented observational evidence that he suggested was consistent with the presence of lichens on Mars.

Carl Otto Lampland: American astronomer who worked at Lowell Observatory from 1901 until 1951; pioneer in photography of planets, whose 1905 photographs of Mars were used by Lowell to confirm to the world the existence of Martians.

Emmanuel Lellouch: French astronomer who led the Infrared Space Observatory scientific team that reported in 2000 that they were unable to detect evidence of methane in the atmosphere of Mars and that the possible detection of methane reported by Krasnopolsky and Mumma in 1997 was, at best, marginal.

Gilbert Levin: team leader of Label-Release experiment on the Viking landers who, with Patricia Straat, continues to assert that this experiment found evidence of life on Mars.

Emmanuel Liais: astronomer at the Paris Observatory and later director of the Observatory of Rio de Janeiro who declared in 1860 that the red color of Mars was due to vegetation.

Percival Lowell: wealthy Boston intellectual who initially built Lowell Observatory in Flagstaff, Arizona, as a facility dedicated to the study of Mars; led a public campaign in which he argued in favor of the existence of Martian engineers who had built a planet-wide system of canals.

Glenn MacPherson: meteorite curator at the Smithsonian Museum of Natural History in Washington, DC, who classified ALH 84001 as a fragment of the asteroid Vesta.

Johann Heinrich von Mädler: German astronomer who, working with Wilhelm Wolff Beer, used observations of Mars made between 1831 and 1839 to create the first global map of Mars.

William C. Maguire: American astronomer who in 1977 determined, using data obtained by Mariner 9 spacecraft in 1972, that the amount of methane in the atmosphere of Mars was below the threshold of detection at that time.

Paul Mahaffy: principal investigator for Sample Analysis at Mars instrument suite on the Mars Curiosity rover who reported, with Christopher Webster, having detected a small amount of methane from November 2013 through January 2014; did not detect methane before (2013 and most of 2014) or after.

Giacomo Filippo Maraldi: the nephew of Cassini and an assistant astronomer at the University of Paris Observatory who in 1704 improved our knowledge of the rotation period of Mars (24h 39m), established that some dark patches on Mars are variable in form and location, and identified bright patches at the north and south poles of Mars that changed in appearance over time.

G. A. Marzo: Italian astronomer who, with S. Fonti, used the Thermal Emission Spectrometer measurements made in 2000–2002 by the Mars Global Surveyor spacecraft to report, in 2010, having detected methane in the Martian atmosphere; claim retracted in 2015.

Chris McKay: American geochemist who, with Rafael Navarro-Gonzalez, carried out experiments with soil from Atacama Desert in Chile that he interpreted as suggestive that the Label-Release experiment on the Viking mission may have detected organic material and possibly life in Martian surface materials.

David McKay: NASA-based scientist who, with Everett Gibson, Jr., Kathie Thomas-Keprta, and Richard Zare, announced in 1996 that ALH 84001 contains fossil evidence of life on Mars that existed nearly 4.1 billion years ago.

Peter Millman: Canadian astronomer who argued that his 1939 observations could neither confirm nor refute the existence of chlorophyll on Mars; also demonstrated that the green color of the supposed Martian seas was inconsistent with the color of terrestrial vegetation.

David Mittlefehldt: American geochemist from Lockheed Engineering who in 1993 concluded that ALH 84001 was a meteorite from Mars.

Michael Mumma: astronomer who, with Vladimir Krasnopolsky, in 1997 reported having detected the possible presence of methane on Mars with data collected with a telescope in Arizona in 1988; also, reported detecting low and variable amounts of Martian methane in 2003 and 2006.

Guido Münch: Mexican astronomer and astrophysicist; *see* Lewis Kaplan.

Bart Nagy: chemist at Fordham University who in 1962 reported finding evidence for extraterrestrial life in the Orgueil meteorite.

Rafael Navarro-Gonzalez: Mexican astronomer who, with Chris McKay, carried out experiments with soil from Atacama Desert in Chile that he interpreted as suggestive that the Label-Release experiment on the Viking mission may have detected organic material and possibly life in Martian surface materials.

Simon Newcomb: a distinguished mathematician and first president of the American Astronomical Society who in 1897 conducted tests of human eyesight to demonstrate that any canals on Mars would be too small to be seen from Earth and that therefore the observations of the supposed Martian canals were impossible.

Vance Oyama: NASA scientist and team leader for Gas-Exchange experiment on Viking landers, which reported no evidence for life on Mars.

Henri Perrotin: French astronomer who in 1886, with Louis Thollon, reported having confirmed the doubling of the Martian canals, first reported by Schiaparelli in 1882.

Edward Charles Pickering: American astronomer and director of the Harvard College Observatory from 1876 until 1920, and older brother of William; in the 1880s raised funds for building and running a high-altitude observatory in Arequipa, Peru.

William Henry Pickering: American astronomer, and younger brother of Edward; while on assignment to Harvard's Boyden observing station in Arequipa, Peru obtained some of the first-ever photographs of Mars, in which he claimed to see both canals and lakes.

George Pimentel: American chemist and team leader for the Mariner 7 infrared spectrometer team who in 1969 first reported having discovered methane in the atmosphere of Mars, almost certainly of biological origin; then retracted this discovery claim when he realized they had discovered frozen carbon dioxide (dry ice) and not methane; *also see* James Shirk.

James Pollack: American planetary scientist who, with Carl Sagan, used one of NASA's deep space network antennas, the Goldstone tracking station in California, to bounce radio waves off the surface of Mars, measure the reflected signals, and interpret the results to explain that the dark areas of Mars change color because they are high-elevation regions that periodically get covered by and then wiped clean of windblown dust.

Richard Anthony Proctor: English astronomer and popularizer of astronomy who in 1867 published *Charts of Mars*, in which he used the Martian drawings made by Dawes to produce a map that included a name—none of which are still in use—for every feature on the surface that he could identify, including seas, oceans, bays, islands, continents, and ice caps; also accurately determined that the rotation period of Mars is 24h 37m 22.7s and wrote an influential essay in *Cornhill* magazine in 1873 arguing that both vegetable life and reasoning creatures must exist on Mars.

Father Giambattista Riccioli: a Jesuit priest who, working with Father Francesco Grimaldi at the Collegio Romano in Rome, reported observing patches on Mars on multiple nights in 1651, 1653, 1655, and 1657.

Carl Sagan: American planetary astronomer involved in Viking missions and the search for large life-forms on Mars; coined the term "macrobe" for life-forms on Mars that might be visible to the naked eye; *also see* James Pollack.

Giovanni Virginio Schiaparelli: Italian astronomer who, after observing Mars intensively in 1877, 1879, and 1881, reported finding more than sixty canals, some thousands of kilometers in length, and that the canal system was regularly expanding through the addition of new canals; also in 1882 he coined the term "gemination" to refer to formation of a second, twin canal, via the sudden appearance of a second canal parallel to a first.

Johann Hieronymus Schröter: German astronomer who, having observed Mars almost continuously for 18 years, from 1785 through 1803, reported that the sizes and shapes of the dark patches on Mars changed regularly and continuously.

Martin Schwarzschild: Princeton University astronomer and director of the Stratoscope II balloon project that in 1963 showed that the amount of carbon dioxide in the atmosphere of Mars was enormous and also that the amount of water vapor was undetectably low and was at best only marginally detected.

Roberta Score: curator of the Antarctic Meteorite Laboratory at the Johnson Space Center in Houston, Texas from 1978 until 1996 who found the meteorite ALH 84001 in Antarctica in December 1984.

Father Angelo Secchi: director of the observatory of the Collegio Romano in Rome, who in 1858 identified two features he called *canali* in his maps of Mars, thereby introducing the word and idea of canals to our studies of Mars; in 1862 he claimed to have proven that continents as well as liquid water and seas exist on Mars; then named many of these seas and oceans and continents and canals for other astronomers.

James Shirk: *see* William Haseltine and George Pimentel.

William Sinton: American astronomer who in the late 1950s identified features in the infrared spectrum of Mars, which became known as the Sinton bands, that he claimed were made by the absorption of light by algae on Mars.

Vesto Melvin Slipher: American astronomer working at Lowell Observatory who in 1908 claimed to have found water in the atmosphere of Mars; in the 1920s he searched for but was unable to find spectral evidence of chlorophyll on Mars.

Hyron Spinrad: American astronomer; *see* Lewis Kaplan.

Patricia Straat: team member of Label-Release experiment on the Viking landers who, with Gilbert Levin, continues to assert that this experiment found evidence of life on Mars.

François Terby: French astronomer who in 1888 reported finding thirty canals on Mars and verified the doubling effect first identified by Schiaparelli in 1882.

Louis Thollon: French astronomer; *see* Henri Perrotin.

Kathie Thomas-Keprta: NASA planetary scientist; *see* David McKay.

Gavriil Adrianovich Tikhov: Russian astronomer who from 1909 through the 1940s searched for proof of the presence of chlorophyll in the colors of reflected light from Mars.

Robert Trumpler: American astronomer who in 1924 provided among the last observational claims confirming the existence of Martian canals; suggested that his work provided evidence for the presence of vegetation along the canal paths.

Geronimo Villanueva: American astronomer who led a team that from 2006 until 2010 searched for evidence of methane on Mars, but found none.

Christopher Webster: Mars Curiosity rover science team member who, with Paul Mahaffy, led investigations for methane on Mars; found no methane in 2012 and most of 2013 and 2014 and 2015, but found evidence of small amounts of methane from November 2013 through January 2014.

Kevin Zahnle: American planetary scientist who has provided alternative explanations to Martian biology for the appearance of methane in measurements by various teams.

Richard Zare: laser chemist at Stanford University; *see* David McKay.

notes

chapter 1. why mars matters

1.   A. P. Nutman, V. C. Bennett, C.R.L. Friend, M. J. Van Kranendonk, and A. R. Chivas, 2016, *Nature*; 537, 535.

2.   NASA press release, 2017, "NASA Affirms Plan for First Mission of SLS, Orion," May 12; https://www.nasa.gov/feature/nasa-affirms-plan-for-first-mission-of-sls-orion

3.   Christian Davenport, 2017, "An Exclusive Look at Jeff Bezos's Plan to Set Up Amazon-Like Delivery for 'Future Human Settlement' of the Moon," *Washington Post*, March 2.

4.   https://www.mars-one.com/about-mars-one

5.   Adam Taylor, 2017, "The UAE's Ambitious Plan to Build a New City—on Mars," *Washington Post*, February 16.

6.   Ishaan Tharoor, 2014, "U.A.E. Plans Arab World's First Mission to Mars, *Washington Post*, July 16.

chapter 2. martians?

1.   http://www.sacred-texts.com/ufo/mars/wow.htm

2.   Richard M. Ketchum, 1989, *The Borrowed Years 1938–1941: America on the Way to War* (New York: Random House), pp. 89–90.

3.   http://www.history.com/this-day-in-history/welles-scares-nation

4.   A. Brad Schwartz, 2015, *Broadcast Hysteria: Orson Welles's "War of the Worlds" and the Art of Fake News* (New York: Hill and Wang), p. 8.

5.   Schwartz, *Broadcast Hysteria*, p. 223.

6.   D. A. Weintraub, 2014, *Religions and Extraterrestrial Life: How Will We Deal With It?* (New York: Springer-Praxis Publishing).

7.   Epicurus, *Letter to Herodotus*. Retrieved from http://www.epicurus.net/en/herodutus .html

8.   S. J. Dick, 1982, *Plurality of Worlds: The Origins of the Extraterrestrial Life Debate from Democritus to Kant* (New York: Cambridge University Press), p. 19.

9.   M. Maimonides, 1986, *Guide for the Perplexed*, as quoted in Norman Lamm, *Faith and Doubt: Studies in Traditional Jewish Thought*, 2nd ed. (New York: KTAV Publishing House), p. 98.

10.   N. Cusanus, 1954, *Of Learned Ignorance*, G. Heron, trans. (New Haven: Yale University Press), pp. 114–115.

11.   G. Bruno, *On the Infinite Universe and Worlds, 1584.* Retrieved from http://www.faculty .umb.edu/gary_zabel/Courses/ParallelUniverses/Texts/OntheInfiniteUniverseandWorlds.htm

12.   Ingrid D. Rowland, 2008, *Giordano Bruno: Philosopher/Heretic* (Chicago: University of Chicago Press).

13.   David A. Weintraub, 2014, *Religions and Extraterrestrial Life* (New York: Springer), pp. 23–24.

14.   Louis Agassiz, in "Tribune Popular Science," 1874, ed. James Thomas Fields; John Greenleaf Whittier (Boston: H. L. Shepard & Co.).

15.   James Jeans, 1942, "Is There Life on the Other Worlds?" *Science*, 95, 589.

16.   Carl Sagan, 1963, "On the Atmosphere and Clouds of Venus," *La Physique des Planètes: Communications Présentées au Onzieme Colloque International d'Astrophysique tenu a Liège*, pp. 328–330.

17.   The Pioneer Venus results were later confirmed by the IUE telescope: Jean-Loup Bertaux and John T. Clarke, 1989, "Deuterium Content of the Venus Atmosphere," *Nature*, 338, 567.

18.   In Michael J. Crowe, 1986, *The Extraterrestrial Life Debate* (Cambridge: Cambridge University Press).

19.   G. Mitri et al., 2014, "Shape, Topography, Gravity Anomalies and Tidal Deformation of Titan," *Icarus*, 236, 169.

20.   NASA press release 15-188, 2015 (September 15), "Cassini Finds Global Ocean in Saturn's Moon Enceladus."

21.   D. A. Weintraub, 2007, *Is Pluto a Planet?* (Princeton, NJ: Princeton University Press).

22.   D. Rittenhouse (1775, February 24). *An oration delivered February 24, 1775, before the American Philosophical Society* (Philadelphia: John Dunlap), pp. 19–20.

23.   Thomas Paine, 1880, *The Age of Reason* (London: Freethought Publishing Company), p. 38.

24.   Michael J. Crowe, 1986, *The Extraterrestrial Life Debate, 1750–1900* (Mineola, NY: Dover).

25.   Stanford Encyclopedia of Philosophy; http://plato.stanford.edu/entries/whewell/

chapter 3. mars and earth as twins

1.   Camille Flammarion, 1892, *La Planète Mars*, in translation as *Camille Flammarion's The Planet Mars*, William Sheehan, ed., Patrick Moore, trans. (London: Springer, 2015), pp. 6–9.

2.   Ibid., pp. 11–12.

3.   Ibid., p. 14.

4.   Ibid.

5.   Ibid., pp. 15–17.

6.   Ibid., pp. 30–31.

7.   Ibid., pp. 34–38.

8.   William Herschel, *Herschel's Second Memoir*, 1784, reproduced in *Camille Flammarion's The Planet Mars*, pp. 48–53.

9.   Flammarion, *La Planète Mars*, pp. 54–74.

chapter 4. imaginary mars

1.   Beer and Mädler, quoted in Flammarion, *La Planète Mars*, p. 92.

2.   Flammarion, *La Planète Mars*, p. 124.

3.   W. R. Dawes, 1865, "On the Planet Mars," *Monthly Notices of the Royal Astronomical Society* 25, 225–268.

4.   Flammarion, *La Planète Mars*, p. 160.

5.   Ibid., p. 114.

6. W. Noble, 1888, "Richard A. Proctor," *The Observatory*, 11, pp. 366–368.

7. Hugh H. Kieffer, Bruce M. Jakosky, and Conway W. Snyder, 1992, "The Planet Mars: From Antiquity to the Present," in *Mars*, ed. H. H. Kiefer et al. (Tucson: University of Arizona Press), p. 28.

chapter 5. misty mars

1. William Huggins, 1867, "On the Spectrum of Mars, with some Remarks on the Colour of that Planet," *Monthly Notices of the Royal Astronomical Society*, 27, 178.

2. Flammarion, *La Planète Mars*, p. 158.

3. Jules Janssen, 1867, *Comptes rendus*, V. LXIV, p. 1304.

4. https://www.ucolick.org/main/

5. W. W. Campbell, 1894, "The Spectrum of Mars," *Publications of the Astronomical Society of the Pacific*, 6, 228.

6. William Huggins, 1895, "Notes on the Atmospheric Bands in the Spectrum of Mars," *Astrophysical Journal*, 1, 193.

7. William Graves Hoyt, 1980, "Vesto Melvin Slipher 1875–1969, A Biographical Memoir" (Washington, DC: National Academy of Sciences).

8. V. M. Slipher, 1908, "The Spectrum of Mars," *Astrophysical Journal*, 28, 397.

9. W. W. Campbell, 1901, "Water Vapor in the Atmosphere of the Planet Mars," *Science*, 30, 771, 474.

10. W. W. Campbell and Sebastian Albrecht, "On the Spectrum of Mars as Photographed with High Dispersion," 1910, *Astronomical Society of the Pacific*, 22, 87.

11. C. C. Kiess, C. H. Corliss, Harriet K. Kiess, and Edith L. R. Corliss, 1957, "High-Dispersion Spectra of Mars," *Astrophysical Journal*, 126, 579.

12. Carl Sagan, 1961, "The Abundance of Water Vapor on Mars," *Astronomical Journal*, 66, 52.

13. Lewis D. Kaplan, Guido Münch, and Hyron Spinrad, 1964, "An Analysis of the Spectrum of Mars," *Astrophysical Journal*, 139, 1.

14. Staff reporter, 1963, "Lower Life Forms May Be Able to Live in Mars Atmosphere, Balloon Findings Show," *Wall Street Journal*, March 5, p. 11.

15. R. E. Danielson et al., 1964, "Mars Observations from Stratoscope II," *Astronomical Journal*, 69, 344.

16. Ronald A. Schorn, 1971, "The Spectroscopic Search for Water on Mars: A History," in *Planetary Atmospheres*, ed. Carl Sagan et al., IAUS, 40, 223–236.

17. Hugh H. Kieffer, Bruce M. Jakosky, and Conway W. Snyder, 1992, "The Planet Mars: From Antiquity to the Present," in *Mars*, ed. H. H. Kiefer et al. (Tucson: University of Arizona Press), p. 11.

chapter 6. red vegetation and reasoning beings

1. "Life in Mars," 1871, *Cornhill* (May), 23, 137, 576–585.

2. Ibid., p. 581.

3. "The Planet Mars—Is It Inhabited?" 1873, *London Reader* (December 1), pp. 69–70.

4. Ibid., p. 70.

5. "The Planet Mars: An Essay by a Whewellite," 1873, *Cornhill* (July), pp. 88–100.

6. Flammarion, *La Planète Mars*, p. 184.

7. Ibid., p. 186.

8. Camille Flammarion, 1879, "Another World Inhabited Like Our Own," *Scientific American Supplement*, 175, p. 2787 (May 10).

chapter 7. water on mars: the real deal

1.   Wilson, S. A. et al., 2016, "A Cold-Wet Middle-Latitude Environment on Mars During the Hesperian-Amazonian Transition: Evidence from Northern Arabia Valleys and Paleolakes," *Journal of Geophysical Research Planets*, 121, 1667.

2.   David E. Smith, Maria T. Zuber, and Gregory A. Neumann, 2001, "Seasonal Variations of Snow Depth on Mars," *Science*, 294, 2142.

3.   Maria T. Zuber et al., 1998, "Observations of the North Polar Region of Mars from the Mars Orbiter Laser Altimeter," *Science*, 282, 2053.

4.   Jeffrey J. Plaut et al., 2008, "Subsurface Radar Sounding of the South Polar Layered Deposits of Mars," *Science*, 316, 92.

5.   Jeremie Lasue et al., 2013, "Quantitative Assessments of the Martian Hydrosphere," *Space Science Reviews*, 174, 155.

6.   G. L. Villanueva et al., 2015, "Strong Water Isotopic Anomalies in the Martin Atmosphere: Probing Current and Ancient Reservoirs," *Science*, 348, 6231, 218.

7.   Jeremie Lasue et al., 2013, "Quantitative Assessments of the Martian Hydrosphere," *Space Science Reviews*, 174, 155.

8.   "Glacial Lake Missoula and the Ice Age Floods," Montana Natural History Center, www .glaciallakemissoula.org

9.   C. M. Stuurman et al., 2016, "SHARAD detection and characterization of subsurface water ice deposits in Utopia Planitia, Mars." *Geophysical Research Letters*, 43, 9484.

10.   NASA press release, 2015, "NASA Mission Reveals Rate of Solar Wind Stripping Martian Atmosphere," November 5.

11.   NASA press release, 2016, "NASA's MAVEN Mission Observes Ups and Downs of Water Escape from Mars," October 19, https://mars.nasa.gov/news/2016/nasas-maven-mission -observes-ups-and-downs-of-water-escape-from-mars

12.   B. M. Jakosky, 2017, "Mars' atmospheric history derived from upper-atmosphere measurements of $^{38}Ar/^{36}Ar$," *Science*, 355, 1408.

13.   NASA press release, 2002, "Found It! Ice on Mars," May 28. https://science.nasa.gov /science-news/science-at-nasa/2002/28may_marsice

14.   W. C. Feldman et al., 2004, "Global Distribution of Near-Surface Hydrogen on Mars," *Journal of Geophysical Research*, 109, E09006.

15.   Roger J. Phillips et al., 2011, "Massive CO_2 Ice Deposits Sequestered in the South Polar Layered Deposits of Mars," *Science*, 332, 838; C. J. Bierson et al., 2016, "Stratigraphy and Evolution of the Buried CO_2 Deposit in the Martian South Polar Cap," *Geophysical Research Letters*, 43, 4172.

16.   P. R. Christensen et al., 2000, "Detection of Crystalline Hematite Mineralization on Mars by the Thermal Emission Spectrometer: Evidence for Near-Surface Water," *Journal of Geophysical Research*, 105, 9623.

chapter 8. canal builders

1.   Flammarion, *La Planète Mars*, p. 251.

2.   Ibid., pp. 300–301.

3.   Ibid., p. 310.

4.   *The Astronomical Register: A Medium of Communication for Amateur Observers*, 236, August 1882, "The Late C. E. Burton," p. 173.

5.   Flammarion, *La Planète Mars*, pp. 333–334.

6.   F. Terby, 1892, "Physical Observations of Mars," *Astronomy and Astro-Physics* (trans. Roger Sprague), 11, pp. 555–558.

7.   E. P. Martz, Jr., 1938, "Professor William Henry Pickering 1858–1938 An Appreciation," *Popular Astronomy*, 46, p. 299 (June–July).

8. William H. Pickering, 1890, "Visual Observation of the Surface of Mars," *Sidereal Messenger*, 9, pp. 369–370.

9. William H. Pickering, 1892, "Mars," *Astronomy and Astro-Physics*, 11, 849.

10. Giovanni Schiaparelli, "The Planet Mars," p. 719, quoted in William Sheehan and Stephen James O'Meara, 2001, *Mars: The Lure of the Red Planet* (Amherst, NY: Prometheus Books), p. 122.

11. Flammarion, *La Planète Mars*, p. 512.

12. William Graves Hoyt, 1976, *Lowell and Mars* (Tucson: University of Arizona Press), pp. 57–58.

13. Ibid., p. 64.

14. Leo Brenner, 1896, "The Canals of Mars Observed at Manora Observatory," *Journal of the British Astronomical Association*, 7, pp. 71–72.

15. Thomas A. Dobbins and William Sheehan, 2007, "Leo Brenner," in *Biographical Encyclopedia of Astronomers*, ed. Virginia Trimble et al. (New York: Springer-Verlag), p. 169.

16. C. A. Young, 1896, "Is *Mars* Inhabited?" *Boston Herald*, October 18 (reprinted in *Publications of the Astronomical Society of the Pacific*, 8, 306, December 1896).

17. Hoyt, *Lowell and Mars*, p. 109.

18. Ibid., p. 124.

19. Ibid., pp. 129–131.

20. Ibid., p. 155.

21. Ibid., p. 163.

22. "Mars," 1907, *Wall Street Journal* (December 28), p. 1.

23. Percival Lowell, 1907, "Mars in 1907," *Nature*, 76, 446.

24. Hoyt, *Lowell and Mars*, p. 141.

25. P. Lowell, 1907, "On a General Method for Evaluating the Surface-Temperature of the Planets; with a Special Reference to the Temperature of Mars," *Philosophical Magazine and Journal of Science*, 14, 79, 161.

26. J. H. Poynting, 1907, "On Professor Lowell's Method for Evaluating the Surface Temperatures of the Planets; with an Attempt to Represent the Effect of Day and Night on the Temperature of the Earth," *Philosophical Magazine and Journal of Science*, 14, 84, 749.

27. Arvydas Kliore, Dan L. Cain, Gerald S. Levy, Von R. Eshleman, Gunnar Fjeldbo, and Frank Drake, 1965, "Occultation Experiment: Results of the First Direct Measurement of Mars's Atmosphere and Ionosphere," *Science*, 149, 1243.

28. Hoyt, *Lowell and Mars*, p. 81.

29. David Strauss, 2001, *Percival Lowell: The Culture and Science of a Boston Brahmin* (Boston: Harvard University Press), p. 230.

30. E. E. Barnard, 1896, "Physical Features of Mars, as Seen with the 36-Inch Refractor of the Lick Observatory, 1894," *Monthly Notices of the Royal Astronomical Society*, 56, 166.

31. Percival Lowell, 1906, "First Photographs of the Canals of Mars," in *Proceedings of the Royal Society of London*, 77, 132.

32. Percival Lowell, 1906, *Mars and Its Canals* (New York: MacMillan), p. 277.

33. Hoyt, *Lowell and Mars*, p. 182.

34. Ibid.

35. Ibid., p. 198.

36. Strauss, *Percival Lowell*, pp. 230–232.

37. Simon Newcomb, 1897, "The Problems of Astronomy," *Science*, 5, 125, 777.

38. Simon Newcomb, 1907, "The Optical and Psychological Principles Involved in the Interpretation of the So-Called Canals of Mars," *Astrophysical Journal*, 26, 1, 1–17.

39. E. M. Antoniadi, 1898, "Chart of Mars in 1896–1897, Considerations on the Physical Condition of Mars, Indistinct Vision and Gemination," *Memoirs of the British Astronomical Association*, 6, pp. 99–102.

40. William Sheehan, 1996, *The Planet Mars* (Tucson: University of Arizona Press), pp. 135–137.

41.   E. M. Antoniadi, 1903, "Report of the Mars Section," *Memoirs of the British Astronomical Association*, 11, pp. 137–142.

42.   E. M. Antoniadi, 1901, "Chart of Mars in 1897–1897, "Chart of Mars in 1898–1899: Conclusion," *Memoirs of the British Astronomical Association*, 9, pp. 103–106.

43.   Sheehan, 1996, *Planet Mars*, p. 140.

44.   E. M. Antoniadi, 1910, "Sixth Interim Report for 1909, Dealing with Some Further Notes on the So-Called 'Canals,'"*Journal of the British Astronomical Association*, 20, 189.

45.   E. M. Antoniadi, 1910, "Considerations of the Physical Appearance of the Planet Mars," *Popular Astronomy*, 21, 416.

46.   Robert Trumpler, 1924, "Visual and Photographic Observations of Mars," *Publications of the Astronomical Society of the Pacific*, 36, 263.

chapter 9. chlorophyll, lichens, and algae

1.   Danielle Briot, 2013, "The Creator of Astrobotany, Gavriil Adrianovich Tikhov," in *Astrobiology, History, and Society*, ed. Douglas A. Vakoch (Heidelberg: Springer), pp. 175–185.

2.   W. W. Coblentz, 1925, "Measurements of the Temperature of Mars," *Scientific Monthly*, 21, 4, pp. 400–404.

3.   V. M. Slipher, 1924, "II. Spectrum Observations of Mars," *Astronomical Society of the Pacific*, 36, 261.

4.   Robert J. Trumpler, 1927, "Mars' Canals Not Man-Made," *Science News-Letter*, 12, 99.

5.   James Stokely, 1926, "Vegetation on Mars?," *Science News-Letter*, 10, 288, 37.

6.   Peter M. Millman, 1939, "Is There Vegetation on Mars?" *The Sky*, 3, 10.

7.   *Life* magazine, 1948 (June 28), "Mars in Color," p. 65.

8.   *Time*, 1948, "The Far-Away Lichens" (March 1).

9.   O. B. Lloyd, 1948, "Astronomers Find Evidence of Life of Primitive Form in Study of Mars," *Toledo Blade* (February 18).

10.   S. Byrne and A. Ingersoll, 2003, "A Sublimation Model for Martian South Polar Ice Features," *Science*, 299, 1051.

11.   Gerard P. Kuiper, 1951, "Planetary Atmospheres and Their Origin," in *The Atmospheres of the Earth and Planets* (Chicago: University of Chicago Press).

12.   Gerard P. Kuiper, 1955, "On the Martian Surface Features," *Publications of the Astronomical Society of the Pacific*, 67, 271.

13.   Gerard P. Kuiper, 1957, "Visual Observations of Mars, 1956," *Astrophysical Journal*, 125, 307.

14.   William M. Sinton, 1958, "Spectroscopic Evidence of Vegetation on Mars," *Publications of the Astronomical Society of the Pacific*, 70, 50.

15.   William M. Sinton, 1957, "Spectroscopic Evidence for Vegetation on Mars," *Astrophysical Journal*, 126, 231.

16.   William M. Sinton, 1959, "Further Evidence of Vegetation on Mars," *Science*, 130, 1234.

17.   N. B. Colthup and William M. Sinton, 1961, "Identification of Aldehyde in Mars Vegetation Regions," *Science*, 134, 529.

18.   Ibid.

19.   Ibid.

20.   D. G. Rea, 1962, "Molecular Spectroscopy of Planetary Atmospheres," *Space Science Review*, 1, 159.

21.   Rea, Belsky, and Calvin, "Interpretation of the 3- to 4-Micron Infrared Spectrum."

22.   James S. Shirk, William A. Haseltine, and George C. Pimentel, 1965, "Sinton Bands: Evidence for Deuterated Water on Mars," *Science*, 147, 48.

23.   D. G. Rea, B. T. O'Leary, and W. M. Sinton, 1965, "Mars: The Origin of the 3.58- and 3.69-Micron Minima in the Infrared Spectra," *Science*, 147, 1286.

24. Ernst J. Öpik, 1966, "The Martian Surface," *Science*, 153, 255.

25. James B. Pollack and Carl Sagan, 1967, "Secular Changes and Dark-Area Regeneration on Mars," *Icarus*, 6, 434.

26. Carl Sagan and James B. Pollack, 1969, "Windblown Dust on Mars," *Nature*, 223, 791.

chapter 10. vikings on the plains of chryse and utopia

1. Tobias Owen et al., 1977, "The Composition of the Atmosphere at the Surface of Mars," *Journal of Geophysical Research*, 82, 4635.

2. Paul R. Mahaffey et al., 2013, "Abundance and Isotopic Composition of Gases in the Martian Atmosphere from the Curiosity Rover," *Science*, 341, 263.

3. Heather B. Franz et al., 2017, "Initial SAM Calibration Experiments on Mars: Quadrapole Mass Spectrometer Results and Implications," *Planetary and Space Science*, 138, 44.

4. Henry S. F. Cooper, Jr., 1980, *The Search for Life on Mars: Evolution of an Idea* (New York: Holt, Rinehart and Winston), p. 68.

5. Ibid., pp. 130–132.

6. John Noble Wilford, 1976, "Viking Finds Mars Oxygen is Unexpectedly Abundant," *New York Times* (August 1).

7. Victor K. McElheny, 1976, "Tests by Viking Strengthen Hint of Life on Mars," *New York Times* (August 8).

8. Victor K. McElheny, 1976, "Mars Life Theory Receives Set Back," *New York Times* (August 11).

9. Victor K. McElheny, 1976, "Tests Continuing for Life on Mars," *New York Times* (August 21).

10. Harold P. Klein et al., 1992, "The Search for Extant Life on Mars," in *Mars*, ed. Hugh H. Kieffer et al. (Tucson: University of Arizona Press), p. 1221.

11. Klein et al., "Search for Extant Life on Mars," p. 1230.

12. Cooper, *Search for Life on Mars*, p. 133.

13. Klein et al., "Search for Extant Life on Mars," p. 1227.

14. Ibid., p. 1230.

15. Gilbert V. Levin, 2015, http://www.gillevin.com/mars.htm

16. G. V. Levin and P. A. Straat, 1979, "Viking Labeled Release Biology Experiment: Interim Results," *Science*, 194, 1322.

17. G. V. Levin and P. A. Straat, 1988, "A Reappraisal of Life on Mars," in *The NASA Mars Conference, Science and Technology Series* 71 (ed. Duke B. Reiber), pp. 186–210.

18. R. Navarro-Gonzalez et al., 2010, "Reanalysis of the Viking Results Suggests Perchlorate and Organics at Midlatitudes on Mars," *Journal of Geophysical Research*, 115, E12010.

19. M. H. Hecht et al., 2009, "Detection of Perchlorate and the Soluble Chemistry of Martian Soil at the Phoenix Lander Site," *Science*, 325, 64.

20. Mike Wall, 2011 (January 6), "Life's Building Blocks May Have Been Found on Mars, Research Finds," http://www.space.com/10418-life-building-blocks-mars-research-finds.html

chapter 11. hot potato

1. Kathy Sawyer, 2006, *The Rock from Mars: A Detective Story on Two Planets* (New York: Random House).

2. I. Weber et al., 2015, *Meteoritics & Planetary Science*, doi: 10.1111/maps.12586.

3. David W. Mittlefehldt, 1994, "ALH 84001, a Cumulate Orthopyroxenite Member of the Martian Meteorite Clan," *Meteoritics*, 29, 214.

4. R. N. Clayton, 1993, "Oxygen Isotope Analysis," *Antarctic Meteorite Newsletter*, 16(3), ed. R. Score and M. Lindstrom (Houston, TX: Johnson Space Center), p. 4.

5.   T. Owen et al., 1977, "The Composition of the Atmosphere at the Surface of Mars," *Journal of Geophysical Research*, 82, 4635.

6.   R. O. Pepin, 1985, "Evidence of Martian Origins," *Nature*, 317, 473.

7.   T. L. Lapen et al., 2010, "A Younger Age for ALH84001 and Its Geochemical Link to Shergotite Sources in Mars," *Science*, 328, 346.

8.   D. S. McKay et al., 1996, "Search for Past Life on Mars: Possible Relic Biogenic Activity in Martian Meteorite ALH 84001," *Science*, 273, 924.

9.   William Clinton, 1996, "President Clinton Statement Regarding Mars Meteorite Discovery," August 7, http://www2.jpl.nasa.gov/snc/clinton.html

10.   Louis Pasteur, 1864 (April 7), "On Spontaneous Generation," speech to Sorbonne.

11.   Johan August Strindberg, 1887, *The Father*, in *Strindberg: Five Plays, 1983*, trans. Harry G. Carlson (Berkeley: University of California Press).

12.   B. Nagy, G. Claus, and D. J. Hennessy, 1962, "Organic Particles Embedded in Minerals in the Orgueil and Ivuna Carbonaceous Chondrites," *Nature*, 4821, 1129

13.   E. Anders et al., 1964, "Contaminated Meteorite," *Science*, 146, 1157.

14.   J. Martel et al., 2012, "Biomimetic Properties of Minerals and the Search for Life in the Martian Meteorite ALH 84001," *Annual Review of Earth and Planetary Sciences*, 40, 167.

15.   Kathy Sawyer, 2006, *The Rock from Mars* (New York: Random House), p. 158.

16.   D. S. McKay et al., 1996, "Search for Past Life on Mars: Possible Relic Biogenic Activity in Martian Meteorite ALH 84001," *Science*, 273, 924.

17.   A. Knoll et al., 1999, *Size Limits of Very Small Microorganisms: Proceedings of a Small Workshop* (Washington, DC: National Academy Press). http://www.nap.edu/read/9638/chapter/1

18.   John D. Young and Jan Martel, 2010, "The Rise and Fall of Nanobacteria," *Scientific American*, 302, pp. 52–59 (January).

19.   J. Martel et al., "Biomimetic Properties of Minerals," p. 183.

20.   Ibid., p. 169.

21.   Ralph P. Harvey and Harry Y. McSween, Jr., 1996, "A Possible High-Temperature Origin for the Carbonates in the Martian Meteorite ALH 84001," *Nature*, 382, 49.

22.   Laurie A. Leshin et al., 1998, "Oxygen Isotopic Constraints on the Genesis of Carbonates from Martian Meteorite ALH 84001," *Geochimica et Cosmochimica Acta*, 62, 3.

23.   Edward R. D. Scott et al., 2005, "Petrological Evidence for Shock Melting of Carbonates in the Martian Meteorite ALH 84001," *Nature*, 387, 377.

24.   J. Martel et al., "Biomimetic Properties of Minerals," p. 175.

25.   Ibid., p. 171.

26.   Ibid., p. 172.

27.   Allan H. Treiman, "Traces of Ancient Life in Meteorite ALH 84001: An Outline of Status in 2003," http://planetaryprotection.nasa.gov/summary/ALH 84001

28.   J. Martel et al., "Biomimetic Properties of Minerals," p. 187.

chapter 12. methane and mars

1.   "Pluto's Methane Snowcaps on the Edge of Darkness," NASA press release, August 31, 2016, https://www.nasa.gov/feature/pluto-s-methane-snowcaps-on-the-edge-of-darkness

2.   R. A. Rasmussen and M.A.K. Khalil, 1983, "Global Methane Production by Termites," *Nature*, 301, 700.

3.   U.S. Environmental Protection Agency, 2016, "Inventory of U.S. Greenhouse Gas Emissions and Sinks: 1990–2014," EPA 430-R-16-002.

4.   G. L. Villanueva et al., 2013, "A Sensitive Search For Organics (CH_4, CH_3OH, H_2CO, C_2H_6, C_2H_2, C_2H_4), Hydroperoxyl (HO_2), Nitrogen Compounds (N_2), NH_3, HCN) and Chlorine Species (HCl, CH_3Cl) on Mars Using Ground-Based High-Resolution Infrared Spectroscopy," *Icarus*, 223, 11.

5.   S. K. Atreya, P. R. Mahaffy, and A.-S. Wong, 2007, "Methane and Related Trace Species on Mars: Origin, Loss, Implications for Life, and Habitability," *Planetary and Space Science*, 55, 358.

6.   Staff reporter, 1966, "Light Wave Study Revives Hope of Martian Life," *New York Times*, October 18, p. 17.

7.   I. S. Bengelsdorf, 1966, "New Analyses May Indicate Biological Activity on Mars," *Los Angeles Times*, October 19, p. 3.

8.   W. Sullivan, 1967, "New Readings on Life on Mars," *New York Times*, February 12, p. 182.

9.   J. Connes, P. Connes, and L. D. Kaplan, 1966, "Mars: New Absorption Bands in the Spectrum," *Science*, 153, 739.

10.   L. D. Kaplan, J. Connes, and P. Connes, 1969, "Carbon Monoxide in the Martian Atmosphere," *Astrophysical Journal*, 157, L187.

11.   W. Sullivan, 1969, "2 Gases Associated with Life Found on Mars Near Polar Cap," *New York Times*, August 8, p. 1.

12.   R. Dighton, 1969, "Mariner Hints Life on Mars," *Atlanta Constitution*, August 8, p. 1A.

13.   R. Abramson, 1969, "New Findings Dim Possibility of Mars Life," *Los Angeles Times*, September 12, p. 3.

14.   D. Horn et al., 1972, "The Composition of the Martian Atmosphere: Minor Constituents," *Icarus*, 16, 543.

15.   Rudy Abramson, 1969, "New Findings Dim Possibility of Mars Life," *Los Angeles Times*, September 12, p. 3; Staff Reporter, 1969, "Unlikelihood of Life on Mars Is Confirmed by Further Study of Mariner 6 and 7 Data," *Wall Street Journal*, September 12, p. 8.

16.   John Noble Wilford, 1972, "Data on Mars Indicate It's a Dynamic Planet; Mars Data Depict a Dynamic Planet That Water Helped Mold; Life Forms Hinted," *New York Times*, June 15, p. 1.

17.   William C. Maguire, 1977, "Martian Isotopic Ratios and Upper Limits for Possible Minor Constituents as Derived from Mariner 9 Infrared Spectrometer Data," *Icarus*, 32, 85.

chapter 13. digging in the noise

1.   V. A. Krasnopolsky, G. L. Bjoraker, M. J. Mumma, and D. E. Jennings, 1997, "High-Resolution Spectroscopy of Mars at 3.7 and 8 μm: A Sensitive Search for H_2O_2, H_2CO, HCl, and CH_4, and Detection of HDO," *Journal of Geophysical Research*, 102, 6525.

2.   E. Lellouch et al., 2000, "The 2.4-45 μm Spectrum of Mars Observed with the Infrared Space Observatory," *Planetary and Space Science*, 48, 1393.

3.   V. A. Krasnopolsky, J. P. Maillard, and T. C. Owen, 2004, "Detection of Methane in the Martian Atmosphere: Evidence for Life?" *Geophysical Research Abstracts*, 6, 06169.

4.   Vladimir A. Krasnopolsky, Jean Pierre Maillard, and Tobias C. Owen, 2004, "Detection of Methane in the Martian Atmosphere: Evidence for Life?" *Icarus*, 172, 537.

5.   "Mars Express Confirms Methane in the Martian Atmosphere," ESA press release, March 30, 2004.

6.   V. Formisano et al., 2004, "Detection of Methane in the Atmosphere of Mars," *Science*, 306, 1758.

7.   A. Geminale, V. Formisano, and M. Giuranna, 2008, "Methane in Martian Atmosphere: Average Spatial, Diurnal, and Seasonal Behaviour," *Planetary and Space Science*, 56, 1194.

chapter 14. here today, gone tomorrow

1.   M. J. Mumma et al., 2003, "A Sensitive Search for Methane on Mars," *Bulletin of the American Astronomical Society*, 35, 937.

2.   M. J. Mumma et al., 2004, "Detection and Mapping of Methane and Water on Mars," *Bulletin of the American Astronomical Society*, 36, 1127.

3. CNN.com, 2004, "Mars Methane from Biology or Geology?," March 30.

4. David A. Weintraub, 2011, *How Old Is the Universe?* (Princeton, NJ: Princeton University Press).

5. M. Peplow, 2004, "Martian Methane Hints at Oases of Life," September 21, http://www.nature.com/news/2004/040920/full/news040920-5.html

6. M. J. Mumma et al., 2005, "Absolute Abundance of Methane and Water on Mars: Spatial Maps," *Bulletin of the American Astronomical Society*, 37, 669.

7. D. J. Harland, 2005, *Water and the Search for Life on Mars* (Chichester, UK: Springer), p. 226.

8. M. J. Mumma et al., 2007, "Absolute Measurements of Methane on Mars: The Current Status," *Bulletin of the American Astronomical Society*, 39, 471.

9. M. J. Mumma et al., 2009, "Strong Release of Methane on Mars in Northern Summer 2003," *Science*, 323, 1041.

10. K. Chang, 2009, "Paper Details Sites on Mars with Plumes of Methane," *New York Times*, January 16.

11. Mumma et al., "Strong Release of Methane on Mars."

12. A. Geminale, V. Formisano, and G. Sindoni, 2011, "Mapping Methane in Martian Atmosphere with PFS-MEX Data," *Planetary and Space Science*, 59, 137.

13. V. A. Krasnopolsky, 2007, "Long-term Spectroscopic Observations of Mars Using IRTF/CSHELL: Mapping of O_2 Dayglow, CO, and Search for CH_4," *Icarus*, 190, 93–102.

14. V. A. Krasnopolsky, 2012, "Search for Methane and Upper Limits to Ethane and SO_2 on Mars," *Icarus*, 217, 144.

15. S. Fonti and G. A. Marzo, 2010, "Mapping the Methane on Mars," *Astronomy & Astrophysics*, 512, A51.

16. S. Fonti et al., 2015, "Revisiting the Identification of Methane on Mars Using TES Data," *Astronomy & Astrophysics*, 581, A136.

17. G. L. Villanueva et al. 2013, "A Sensitive Search for Organics."

18. Ibid.

chapter 15. sniffing martian air with curiosity

1. http://www.robothalloffame.org/inductees/03inductees/mars.html

2. C. R. Webster et al., 2013, "Low Upper Limit to Methane Abundance on Mars," *Science*, 342, 355.

3. Ibid.

4. C. R. Webster et al., 2015, "Mars Methane Detection and Variability at Gale Crater," *Science*, 347, 415.

5. C. Sagan, 1998, *Billions and Billions: Thoughts on Life and Death at the Brink of the Millennium* (New York: Ballantine), pp. 60 and 85.

6. K. Zahnle, R. S. Freedman, and D. C. Catlin, 2011, "Is There Methane on Mars?," *Icarus*, 212, 493.

7. https://www.nasa.gov/feature/goddard/2016/maven-observes-ups-and-downs-of-water-escape-from-mars

8. J. A. Holmes, S. R. Lewis, and M. R. Patel, 2015, "Analysing the Consistency of Martian Methane Observations by Investigation of Global Methane Transport," *Icarus*, 257, 32.

9. J. Bontemps, 2015, "Mystery Methane on Mars: The Saga Continues," *Astrobiology Magazine*, May 14, http://www.astrobio.net/news-exclusive/mystery-methane-on-mars-the-saga-continues/

10. K. Zahnle, 2015, "Play It Again, SAM," *Science*, 347, 370.

11. Ibid., p. 371.

12.   C. Oze and M. Sharma, 2005, "Have Olivine, Will Gas: Serpentinization and Abiogenic Production of Methane on Mars," *Geophysical Research Letters*, 32, L10203.

13.   S. K. Atreya, P. R. Mahaffy, and A.-S. Wong, 2007, "Methane and Related Trace Species on Mars: Origin, Loss, Implications for Life, and Habitability," *Planetary and Space Science*, 55, 358.

14.   A. Bar-Nun and V. Dimitrov, 2006, "Methane on Mars: A Product of H_2O Photolysis in the Presence of CO," *Icarus*, 181, 320–322, and 2007, "'Methane on Mars: A Product of H_2O Photolysis in the Presence of CO' Response to V. A. Krasnopolsky," *Icarus*, 188, 543.

15.   F. Keppler et al., 2012, "Ultraviolet-Radiation-Induced Methane Emissions from Meteorites and the Martian Atmosphere," *Nature*, 486, 93.

16.   B. K. Chastain and V. Chevrier, 2007, "Methane Clathrate Hydrates as a Potential Source for Martian Atmospheric Methane," *Planetary and Space Science*, 55, 1246.

17.   NASA press release, May 11, 2016, "Second Cycle of Martian Seasons Completing for Curiosity Rover," https://mars.nasa.gov/news/second-cycle-of-martian-seasons-completing-for-curiosity-rover

chapter 16. chasing martians

1.   Raymond E. Arvidson, 2016, "Aqueous History of Mars, as Inferred from Landed Mission Measurements of Rocks, Soils and Water Ice," *Journal of Geophysical Research Planets*, 121, 1602.

2.   www.unoosa.org/oosa/en/ourwork/spacelaw/treaties/introouterspacetreaty.html

3.   https://cosparhq.cnes.fr

4.   Christopher P. McKay, 2007, "Hard Life for Microbes and Humans on the Red Planet," *AdAstra*, 31.

5.   Carl Sagan, 1980, *Cosmos* (New York: Random House), ch. 5.

glossary

absorption band: a section of the spectrum of light from a source from which some or all of the light is missing; caused by the material located in between the observer and the original light source that absorbs light at very distinct colors or wavelengths.

albedo: the proportion of incident sunlight that is reflected by the surface of a planet or moon.

algae: aquatic, photosynthetic organisms that can live in freshwater or seawater; lack roots, stems, and leaves; can exist as single, microscopic cells or as multicellular, macroscopic organisms.

ALH 84001: (Allan Hills 84001), a meteorite from Mars, collected by Roberta Score in Antarctica in 1984, that may show fossil evidence for ancient life on Mars.

ancient valley networks: features on the ancient Martian surface, some as much as 0.6 miles (1 kilometer) in width and 600 feet (several hundred meters) in depth, that resemble river valley networks on Earth.

areography: term, akin to geography, coined in the nineteenth century for the mapping of the surface of Mars (after Ares, Greek god of war).

areology: the study of the planet Mars.

canale: Italian word (singular) for channel; can be used to refer to a canal constructed by human engineers, a body of water as broad and deep as the English Channel, or a mountain gully; first applied to Mars in 1858 by Father Angelo Secchi and made popular in the 1880s in Giovanni Schiaparelli's maps of Mars (plural, canali).

Cepheid: a star whose brightness increases and decreases periodically, and whose period of brightness changes is directly correlated with the maximum brightness of the star, such that the brightest Cepheids have long periods (about 100 days) and the faintest Cepheids have periods of less than a day; used by astronomers to determine distances to clusters of stars and galaxies that have within them at least one Cepheid.

chlorophyll: a molecule in plants that absorbs light in the process of photosynthesis; by absorbing most colors of visible light except green, which is reflected, chlorophyll gives plants their green color.

deuterium: a heavy hydrogen atom, containing both a proton and a neutron in the nucleus, whereas normal hydrogen has only a single proton in the nucleus.

diogenite: a meteorite that formed as igneous rock that cooled slowly, deep inside a large parent-body, such as a moon, planet, or asteroid.

gemination: the process by which, according to the claims of Giovanni Schiaparelli, the supposed canals on Mars suddenly changed from single canals to double canals.

heavy water: a water molecule (H_2O) in which both hydrogen atoms have been replaced with deuterium atoms (thus, D_2O); semi-heavy water would have the form HDO.

lichen: long cells of fungi, many of which contain multiple nuclei, connected end-to-end to form long, tubular filaments with cell walls receiving structural support from chitin, a carbohydrate polymer molecule; live symbiotically with photosynthetic cells, usually green algae but sometimes an ancient form of bacteria called cyanobacteria.

lithophile: an element that bonds easily and readily with oxygen, thereby forming oxides and silicates; found concentrated in the mantle and crust (rather than the core) of Earth.

macrobe: word coined by Carl Sagan for a hypothesized life-form on Mars (or possibly elsewhere in the universe) that is large enough to be obvious in photographs taken from a camera (rather than a microscope or other specialized detection equipment) on the surface of a planet.

metaphysics: an ancient branch of philosophy that attempts to explain everything about the world (motion, space, time, substance, existence) on the basis of fundamental principles determined by rational thought, rather than on the basis of knowledge gained from experiments and observations of the physical universe.

methane: a molecule made up of one carbon atom and four hydrogen atoms (CH_4).

methanogenic bacteria: anaerobic (living in environments absent of free oxygen) bacteria that produce methane gas as a by-product of their energy metabolism.

moss: a small, nonvascular land plant that reproduces with spores rather than flowers and seeds and that absorbs water and nutrients mainly through leaves rather than roots.

opposition: an alignment of two planets on the same side of the Sun, such that the Sun and the two planets form a straight line, and so the second planet, when viewed from Earth, is seen in exactly the opposite direction as the Sun and the full disk of the planet is illuminated by sunlight.

organic molecule: a molecule that contains one or more carbon atoms and carbon-hydrogen (C-H) bonds; molecules associated with all living things on Earth, including DNA, fats, sugars, proteins, and enzymes, are organic molecules.

outflow channels: features on the ancient Martian surface, some as much as tens of miles wide and nearly a thousand miles in length, that appear to have been carved by the quick melting and catastrophic release of enormous volumes of water.

photographic plate: a piece of glass coated with light-sensitive chemicals that, when exposed to light collected by a telescope and then bathed in the right darkroom chemical baths, yields images of celestial objects; used by professional astronomers from the 1890s through the 1970s.

plurality of worlds hypothesis: A medieval and Renaissance-era idea that claimed that every star, planet, or moon in the universe (all of them being "worlds") is inhabited by intelligent beings that are able to worship God.

precipitable water vapor: the total depth of water on the surface of a planet if all of the moisture in the atmosphere condensed onto the surface in liquid form.

principle of plenitude: twentieth-century American intellectual historian Arthur Lovejoy identified the principle of plenitude as the idea that "no genuine potentiality of being can remain unfulfilled, that the extent and abundance of the creation must be as great as the possibility of existence ... that the world is better, the more things it contains."*

semi-heavy water: *see* heavy water.

siderophile: an element that tends not to form bonds with oxygen and sulfur and that is readily soluble in molten iron; siderophiles are concentrated in the core (rather than the mantle or crust) of Earth.

sol: one day (the time from sunrise to the next sunrise) on Mars, equal to 24 hours, 39 minutes, 35.244 seconds; this length of time results from the combination of Mars's actual rotation period and the forward motion of Mars in its orbit around the Sun.

spectroscopy: channeling a beam of light from any source of light through a prism or reflecting the light off a grating, which spreads the light out into its constituent colors, allowing one to study the details of brightness and faintness of the different colors.

spectrum: the detailed rainbow of light obtained by allowing light from a source to pass through a prism or reflect off a grating; the spectrum of a celestial object, in addition to showing light across a breadth of colors, will show some colors brighter and others fainter than average.

terraforming: to change the environment, in particular the contents and temperature of the atmosphere, of another planet into one resembling Earth, such that it becomes habitable for humans.

terrestrial fractionation line: a way of distinguishing material from the Earth-Moon system from material from other planets, that utilizes the way in which isotopes of a single element like oxygen separate (or fractionate) as a result of processes like evaporation or chemical bonding.

---

*Arthur O. Lovejoy, 1971, *The Great Chain of Being: A Study of the History of an Idea* (Cambridge, MA: Harvard University Press).

index